SOLUTIONS MANUAL TO ACCOMPANY

The Engineering Design Process

Atila Ertas
Jesse C. Jones
Department of Mechanical Engineering
Texas Tech University

John Wiley & Sons, Inc.
New York   Chichester   Brisbane   Toronto   Singapore

Copyright © 1993 by John Wiley & Sons, Inc.

This material may be reproduced for testing or instructional purposes by people using the text.

ISBN 0-471-58315-4

Printed in the United States of America

10 9 8 7 6 5 4 3 2 1

Printed and bound by Malloy Lithographing, Inc.

CONTENTS

CHAPTER 1	1
CHAPTER 2	5
CHAPTER 3	9
CHAPTER 4	22
CHAPTER 5	35
CHAPTER 6	47
CHAPTER 7	52
CHAPTER 8	61
CHAPTER 10	72
APPENDIX B	105
SOLUTION TO EXERCISE 1.1	118

0.1 Chapter 1

1.1

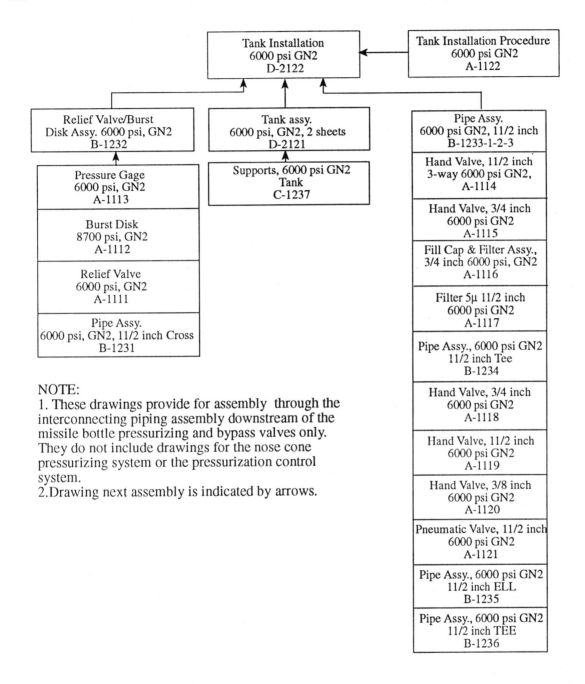

NOTE:
1. These drawings provide for assembly through the interconnecting piping assembly downstream of the missile bottle pressurizing and bypass valves only. They do not include drawings for the nose cone pressurizing system or the pressurization control system.
2. Drawing next assembly is indicated by arrows.

1.2
This is a brainstorming problem for student groups. Several alternative substitutes for windshield wipers should be developed as a result of the brainstorming sessions.

1.3
Several alternative design concepts to the typical camshaft valve mechanism in internal combustion engines should be developed by the student in response to this problem. Solenoid valve actuation

should not be overlooked but there are several problems that must be solved for this to be a viable alternative. The student should recognize these difficulties in his analysis.

1.4

Either methanol or natural gas may be acceptable as a fuel for yard and garden maintenance equipment. The student should develop adequate justification for his selection. One feature that should not be overlooked is the possibility of using a small home compressor connected to the low pressure natural gas supply for the house to supply fuel for yard equipment. This may prove to be very inexpensive as well as convenient in that it eliminates the need to journey to a service station to procure fuel. Both of these alternative fuels are much better than gasoline from an emissions standpoint.

1.5

Up to the first lunar landing in 1969 the space program had a very clear mandate to achieve this significant undertaking. Funding was abundantly available, the general public strongly supported the program, and within NASA the priorities were clear - the Apollo program took precedence over all other activities. With this level of support NASA took the ball and ran with it successfully. The problem was that no significant effort or thought was given to what the agency would do if, and when, they were successful in placing a man on the moon and returning him. After the first lunar landing in June 1969 funding became much more difficult to obtain through the Congress and the general public lost interest in missions that seemed to be only repeating what had already been achieved. Worse yet, the agency no longer had a mission that was unquestionably supported and NASA employees began questioning where the agency was going and what could be done to reestablish the support that had existed before the lunar landing. The problem was that NASA had not planned on such an immediate success and were not prepared to cope with it when it occurred. They had a very worthy and strongly supported objective up to and until the successful lunar landing but had no objective once that goal had been achieved. When an organization does not have an objective that is widely and strongly supported it begins to doubt its purpose and question its existence. This is exactly what happened to NASA in late 1969 and through the early 1970's, until a new mission (the space shuttle) and set of objectives was identified and approval was obtained through the Congress. Most of the discussion within NASA during this time period was associated with what could be done to convince the public and Congress of the worthiness of NASA's efforts. The problem was thought to be inability of the agency to sell itself rather than that the agency no longer had an objective that could capture the imagination and support of the general public, Congress and NASA's own employees.

1.6

a. Top mangement must be heavily involved during the early phases of a project since they must approve effort expended on new endeavors. They may even identify the need that is the focus of the effort. They would also normally remain close to the project through the early conceptualization and feasibility phases since the decision to proceed and the funding authorization is made by top management. Top management would also be actively involved in approving (or making) organizational assignments and approving the work breakdown structure. With the start of preliminary design top management would probably revert to a monitoring role and thus not be as actively involved in the project. They will continue to be involved in periodic project reviews and in major project decisions having to do with project objectives and the ability to meet contractual requirements. Top management will again become actively involved in the decision to proceed into production since this is a major commitment on the part of the firm. They will also determine production quantities.

b. Marketing should be consulted at the time the decision is made to proceed with the project since they can provide input as to the sales potential of the product. This organization can also provide customer input that may influence the design of the product so they should also be consulted during preliminary design and early during the detailed design phase of the project. Marketing has an input to production since they are in the best position to assess potential sales and thus provide input for top management to determine production quantities.

c. As indicated in Chapter 1, the trend is for production to become involved early in the design phase of the project to minimize total development time and to lessen the problems associated in going from design to production. If the firm has an integrated design/production organization production engineers will be involved very early in the design effort to insure that the product can be produced cost-effectively and with minimum effort and commitment of time.

The project manager is the person who has primary responsibility for insuring that input is obtained from all appropriate elements of the organization at the proper time and to the required level.

1.7

Design - In accordance with ABET's definition, design includes all activities shown on Figure 1.1 beginning with recognition of need through completion of detailed design and successful qualification testing. Product improvement also involves design effort and thus design often extends over the life of the product.

Development - Development is generally that activity required to support the detailed design phase of the project and is primarily associated with testing and process definition. Development also supports product improvement design efforts and can thus also extend over the life of the product. Development is sometimes used to refer to earlier project effort in which case it is referred to as "basic" development or "advanced" development.

Engineering - Engineering includes all activity related to the product that is performed by engineers over the product's life.

1.8

Development testing - All testing that supports the design and development of the product is considered to be development testing.

Qualification testing - Testing that verifies that the product was fabricated in accordance with the design drawings and specifications and meets the project requirements.

Acceptance testing - Testing that verifies that the production processes and procedures have been correctly applied during fabrication of the product.

In an attempt to reduce overall project costs the major structural test article in the Space Shuttle program was recycled back into the production line after completion of testing and was converted into a production article. This is probably an extreme example of cost-cutting but in a tightly funded program such cost-cutting methods are often considered necessary. A more common way in which project costs are controlled is to maintain close control over development testing since it is a high cost element. Unfortunately, reduced testing often means increased risk, which can place project engineers at odds with project management. Another method of controlling project costs is to minimize the number of qualification test articles and try to maximize the number of systems that can be qualified with a single test article. This also may tend to place project engineering at odds with project management.

1.9

The most obvious place to reduce the time for bringing the product to the marketplace is to integrate detailed design and production planning and tooling design as suggested in Chapter 1. If the project can move almost directly from detailed design into production a significant amount of time can be cut from the overall product development time. Another possibility would be to integrate the conceptualization and feasibility assessment activities. This might be rather easily accomplished but will probably not result in major project schedule compression.

1.10

a. Set up a procedure to maintain close monitoring of the fuel added and extracted from the tank in an attempt to positively determine whether a leak exists.
b. Perform testing in the basements of nearby buildings to determine if fuel residuals can be detected.
c. Drill monitoring wells at appropriate locations and establish methods for taking samples.
d. Take soil and water samples during drilling of the monitoring wells.
e. Take organic vapor readings during drilling of the monitoring wells.
f. Take soil borings at appropriate locations.
g. Analyze all the data and determine whether the tank is leaking and, if so, the extent of the contaminate plume.
h. Develop a plan for removal of the leaking tank.
i. Remove the leaking tank.
j. Develop a plan for site remediation. Emphasis should be on minimum cost that will satisfactorily restore the site. Approaches such as soil venting should be considered to determine whether adequate fuel residuals can be removed with this method. If not, the soil will have to removed and either decontaminated on site using some method such as thermal stripping or be relocated and reburied at a controlled site.

0.2 Chapter 2

2.1
Schedule of Management activities for project planning Exercise 2.2

TASK	1	2	3	4	5	6	7	8	9	10	11	12	13	14	15	16	17	18	19	20
1. Gather and analyze the facts of the current project situation.	☐																			
2. Establish project objectives.		☐																		
3. Develop possible alternative courses of actiona and Identify the negative consequences.			☐																	
4. Decide on a basic course of action.				▽																
5. Identify and analyze the various tasks necesary to implement the project and develop strategies (priorities, sequence, timing of major steps).					☐															
6. Establish qualifications for positions and define scope of relationships, responsibilities and authority of new positions.					☐☐															
7. Establish appropriate policies for recognizing individual performance.							☐☐													
8. Determine the allocation of resources (budget, facilities, etc.).							☐☐☐☐☐☐☐☐☐☐☐☐☐☐													
9. Find quality people to fill positions.									☐☐☐											
10. Train and develop personnel for new responsibilities/authority.											☐☐☐☐☐☐									
11. Develop individual performance objectives which are mutually aggreeable to the individual and supervisor.																		☐☐		
12. Assign responsibility/accountability/authority.																				▽

2.2

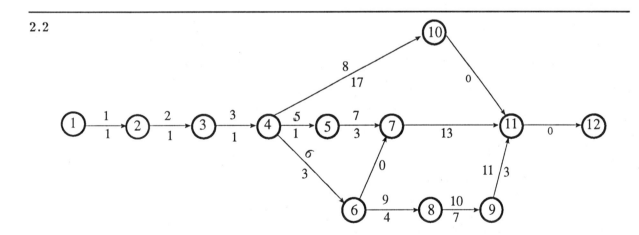

NOTE: Activities are identified by the numbers above arrow. Schedule time is identified by the numbers below the arrow. Events (milestones) are identified by the numbers within circles.

2.3
This a "WHAT" versus "HOW" orthogonal array based on the individual students desires for his first job after graduation.

2.4

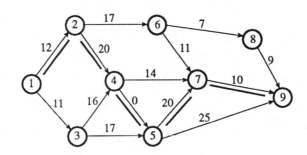

Activities		Estimates			
Successor event	Predecessor event	T_e	T_E	T_L	Slack $T_L - T_E$
9	8	9	45	62	17
9	7	10	62	62	0
9	5	25	57	62	5
8	6	7	36	53	17
7	6	11	40	52	12
7	5	20	52	52	0
7	4	14	46	52	6
6	2	17	29	41	12
5	4	0	32	32	0
5	3	17	28	32	4
4	3	16	27	32	5
4	2	20	32	32	0
3	1	11	11	15	4
2	1	12	12	12	0

2.5

	Conversion cost	Infrastructure cost	Safety	Convenience	Environmental effect	Vehicle Maintenance	US trade balance	Feedstock availability	Energy Security	Fuel cost	Vehicle range	Score
a. CNG	4	4	5	2	5	5	5	5	5	5	2	47
b. LNG	2	1	3	2	5	5	5	5	5	4	4	41
c. MIOO	0	3	4	5	4	4	5	5	5	3	3	41
d. M85	0	3	3	5	3	4	3	3	3	3	3	33
e. E100	0	3	4	5	4	4	5	4	5	1	3	38
f. E85	0	3	3	5	3	4	3	2	3	1	3	32
g. LPG	5	5	3	3	4	5	3	3	3	4	4	42

2.6

	External connection	Connect parts	House parts	Stop flow	Stop leakage	Resist corrosion	Allow flow	Total cost ($)	% Total cost
1. Inlet body	5	4	3	3	1	1	3	20	33
2. O' Ring valve seal				0.5	0.5			1	2
3. Body seal Gasket					5			5	8
4. Poppet				4		2	4	10	16
5. Spring				2	1	1	1	5	8
6. Outlet body	5	5	4	1	4	1		20	33
Total	10	9	7	10.5	11.5	5	8	61	100
%Total	16	15	11	17	19	8	13		
High or low		H			H				

Based on this analysis the ``Stop Leakage'' function might be considered a relatively high cost parameter and other design options could be considered. The inlet and outlet bodies are the high cost components of the valve and methods of reducing their cost such as using lower cost materials could be considered. Overall the valve appears to be pretty well designed.

2.7

Parts \ Failure Modes	Weight	Corrodes	Leaks	Breaks	Galls	Yields	Swells	Gross weight	%
1. Inlet body	10	30	30	90	30			180	25
2. O Ring valve seal	5		45				45	105	15
3. Body seal Gasket	2		18	6				24	3
4. Poppet	7	63		21				84	12
5. Spring	7	21		63	63			147	20
6. Outlet body	10	30	30	90	30			180	25
								720	100

From the FMEA it is clear that the components that should be designed so that failure (and the consequences therefrom) is minimized are the spring, the O' ring valve seal, and the inlet and outlet bodies.

2.8 Loss Function

$$L(Y) = k(Y-m)^2$$
$$k = \frac{L(Y)}{(Y-m)^2} = \frac{150}{(\mp 10)^2}$$
$$k = 1.5$$

For the limit that defines when an adjustment should be made at the factory rather than at the dealer's repair shop assume that the cost to the customer in dollars is equal to the cost to have the adjustment made at the dealer's repair shop. Therefore,

$$\Delta = \sqrt{\frac{A}{A_0}}(\Delta_0)$$
$$= \sqrt{\frac{15}{150}}(\mp 10)$$
$$= \mp 3.16 \text{ cps}$$

0.3 Chapter 3

3.1
Since this problem is open-ended, no solution is provided.

3.2
(a) SINGLE DEGREE-OF-FREEDOM MODEL
Assumptions:
- center of mass coincides with the geometric center, i.e. $l_1 = l_2 = l$.
- stiffness and damping for each tire set are equal, i.e. $k_1 = k_2 = k$ and $c_1 = c_2 = c$.
- no angular oscillations, i.e. $\theta = 0$.

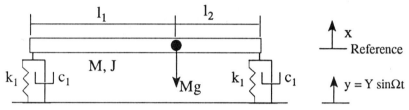

Fig. 3.2.1 model of a car

Where M is the mass of the car and J is the mass moment of inertia about the center of mass.

The freebody diagram of the automobile can be drawn as shown in Fig. 3.2.2:

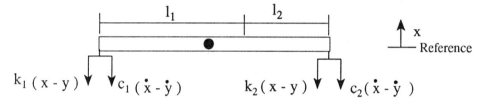

Fig. 3.2.2 Free body diagram

The motion of the center of mass is given by Newton's second law as

$$m\ddot{x} = -k_1(x-y) - k_2(x-y) - c_1(\dot{x}-\dot{y}) - c_2(\dot{x}-\dot{y}) \tag{3.2.1}$$

Let $x - y = z$ and using $k_1 = k_2 = k$ and $c_1 = c_2 = c$, the above equation becomes

$$m\ddot{z} + 2c\dot{z} + 2kz = -m\ddot{y} \tag{3.2.2}$$

Using $y = Y\sin\omega t$ and letting $m\omega^2 Y = F_1$, yields

$$m\ddot{z} + 2c\dot{z} + 2kz = F_1 \sin\omega t \tag{3.2.3}$$

(b) TWO DEGREE-OF-FREEDOM MODEL Assumptions:
- center of mass does not coincide with the geometric center, i.e. $l_1 \neq .l_2$.
- stiffness and damping for each tire set are not equal, i.e. $k_1 \neq .k_2$ and $c_1 \neq .c_2$.
- small angular oscillations, i.e. $\sin\theta \approx \theta$.

The freebody diagram of the automobile can be drawn as shown in Fig. 3.2.3:
The motion of the center of mass is given by Newton's second law as

$$m\ddot{x} = -k_1(x-y-l_1\theta) - k_2(x-y+l_2\theta) - c_1(\dot{x}-\dot{y}-l_1\dot{\theta}) - c_2(\dot{x}-\dot{y}+l_2\dot{\theta}) \tag{3.2.4}$$

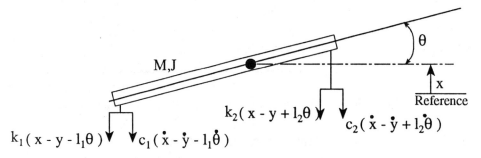

Fig. 3.2.3 Free body diagram

Summing moments about the mass center, the equation governing rotation can be formulated as

$$J\ddot{\theta} = k_1(x - y - l_1\theta)l_1 - k_2(x - y + l_2\theta)l_2 + c_1(\dot{x} - \dot{y} - l_1\dot{\theta})l_1 - c_2(\dot{x} - \dot{y} + l_2\dot{\theta})l_2 \qquad (3.2.5)$$

Let $x - y = z$, the above equations become

$$m\ddot{z} + (c_1 + c_2)\dot{z} - (c_1 l_1 - c_2 l_2)\dot{\theta} + (k_1 + k_2)z - (k_1 l_1 - k_2 l_2)\theta = -m\ddot{y} \qquad (3.2.6)$$

$$J\ddot{\theta} - (c_1 l_1 - c_2 l_2)\dot{z} + (c_1 l_1^2 + c_2 l_2^2)\dot{\theta} - (k_1 l_1 - k_2 l_2)z + (k_1 l_1^2 + k_2 l_2^2)\theta = 0 \qquad (3.2.7)$$

Using $y = Y \sin \omega t$ and letting $m\omega^2 Y = F_1$, yields

$$m\ddot{z} + (c_1 + c_2)\dot{z} - (c_1 l_1 - c_2 l_2)\dot{\theta} + (k_1 + k_2)z - (k_1 l_1 - k_2 l_2)\theta = F_1 \sin \omega t \qquad (3.2.8)$$

$$J\ddot{\theta} - (c_1 l_1 - c_2 l_2)\dot{z} + (c_1 l_1^2 + c_2 l_2^2)\dot{\theta} - (k_1 l_1 - k_2 l_2)z + (k_1 l_1^2 + k_2 l_2^2)\theta = 0 \qquad (3.2.9)$$

3.3

$$\pi_1 = \mu^{x_1} d^{y_1} v^{z_1} h$$

Following Step 3, the above equation can be written in basic units as

$$M^0 L^0 T^0 = (ML^{-1}T^{-1})^{x_1}(L)^{y_1}(LT^{-1})^{z_1}(L)^1$$

hence,

$$\begin{aligned} \text{for } M, \ 0 &= x_1 \\ \text{for } L, \ 0 &= -x_1 + y_1 + z_1 + 1 \\ \text{for } T, \ 0 &= -x_1 - z_1 \end{aligned}$$

Solution of the above equations yields, $x_1=0$, $y_1=-1$, and $z_1=0$. Thus,

$$\pi_1 = \mu^0 d^{-1} v^0 h = d^{-1} h = \frac{h}{d}$$

Table 0.1: Physical characteristics [hp]

Physical characteristic	Symbol	Basic dimension
Distance	h	L
Time	t	T
Gravity	g	LT^{-2}
Weight	W	MLT^{-2}

Similarly,

$$\pi_2 = \frac{vt}{d}, \quad \pi_3 = \frac{dg}{v^2}, \quad \pi_5 = \frac{mv}{\mu d^2}$$

These π terms are unique and independent each other.

3.4

The number of selected physical characteristics, $n=4$, and the repeating basic dimensions are M, L, and time T, therefore, $b=3$. Hence, the number of necessary π terms, S, is

$$\begin{aligned} S &= n - b \\ &= 4 - 3 = 1 \end{aligned}$$

b) The physical characteristics can be written in a functional equation as

$$F(h, t, g, W) = 0$$

and the functional equation can be written in dimensional form as

$$F(L, T, LT^{-2}, MLT^{-2}) = 0$$

The repeating variables are; t g, and W.

$$\pi_1 = t^x g^y W^z D$$

The above equation can be written in basic units as

$$M^0 L^0 T^0 = (T)^x (LT^{-2})^y (MLT^{-2})^z (L)$$

$$\begin{aligned} \text{for } M, \quad 0 &= z \\ \text{for } L, \quad 0 &= y + z + 1 \\ \text{for } T, \quad 0 &= x - 2y - 2z \end{aligned}$$

Solution of the above equations yields, $x=-2$, $y=-1$, and $z=0$. Hence,

Table 0.2: Physical characteristics
[hp]

Physical characteristic	Symbol	Basic dimension
Deflection	y	L
width	d	L
Load location from left	a	L
Coordinate of the deflection	x	L
Load	P	MLT^{-2}

$$\pi = t^{-2}g^{-1}W^0 D = \frac{D}{gt^2}$$

or π equation, $F(\frac{D}{gt^2}) = 0$

In the particular case in which only one π term evolves, the π equation gives as an explicit equation for the desired physical characteristic as

$$D = gt^2$$

Including numerical coefficient yields

$$D = Cgt^2$$

In the case of free fall C=1/2.

3.5

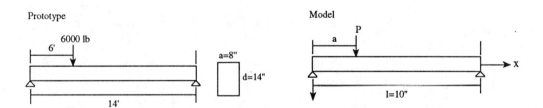

Basic unit for P can also be written as

$$P = \frac{AEy}{l} = \frac{(L)^2(E)(L)}{L} = EL^2$$

Remembering any dimensionless physical characteristics such as $\frac{width}{length}$ ratio may be immediately selected as π terms, thus

$$\pi_1 = \frac{y}{l}, \; \pi_2 = \frac{w}{l}, \; \pi_3 = \frac{d}{l}, \; \pi_4 = \frac{a}{l}, \; \pi_5 = \frac{x}{l}, \; \pi_6 = \frac{p}{El^2}$$

Define the *scale length* between model and prototype such that

$$\text{Scale length} = \frac{\text{prototype length}}{\text{model length}}$$

or

$$\lambda = \frac{l_p}{l_m} = \frac{(14)(12)}{10} = 16.8''$$

$\lambda_2 = \frac{w}{l}$ and $\lambda_3 = \frac{d}{l}$ indicates that the model and prototype will be geometrically similar.

$$\lambda_{2m} = \lambda_{2p} \Rightarrow \frac{w_m}{l_m} = \frac{w_p}{l_p}$$

$$w_m = w_p \frac{l_m}{l_p} = 8\frac{1}{16.8} = 0.476''$$

$$\lambda_{3m} = \lambda_{3p} \Rightarrow \frac{d_m}{l_m} = \frac{d_p}{l_p}$$

$$d_m = d_p \frac{l_m}{l_p} = 14\frac{1}{16.8} = 0.833''$$

$$\lambda_{4m} = \lambda_{4p} \Rightarrow \frac{a_m}{l_m} = \frac{a_p}{l_p}$$

$$a_m = a_p \frac{l_m}{l_p} = (6)(12)\frac{1}{16.8} = 4.286''$$

$$\lambda_{5m} = \lambda_{5p} \Rightarrow \frac{x_m}{l_m} = \frac{x_p}{l_p}$$

$$x_m = x_p \frac{l_m}{l_p} = x_p \frac{1}{16.8}$$

$$\lambda_6 = \frac{P}{El^2}$$

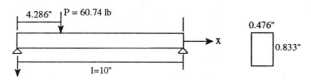

$$\frac{P_m}{E_m l_m^2} = \frac{P_p}{E_p l_p^2} \Rightarrow P_m = P_p \frac{E_m}{E_p} \frac{l_m^2}{l_p^2} = (6000)\frac{30 \times 10^6}{10.5 \times 10^6}(\frac{1}{16.8})^2 = 60.74 \ lb_f$$

From π_1, $y_p = y_m \frac{l_p}{l_m} = 16.8 y_m$

3.6

The number of selected physical characteristics, $n=6$, and the repeating basic dimensions are M, L, and T, therefore, $b=3$. Hence, the number of necessary π terms, S, is

$$\begin{aligned} S &= n - b \\ &= 6 - 3 = 3 \end{aligned}$$

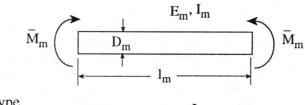

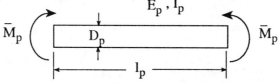

Table 0.3: Physical characteristics

Physical characteristic	Symbol	Basic dimension
Bending moment	M	ML^2T^{-2}
length	l	L
Diameter	D	L
Modulus of elasticity	E	$ML^{-1}T^{-2}$
Area moment of inertia	I	L^4
Density	ρ	ML^{-3}

Rewrite T as

$$T \sim \left[\frac{LT^2}{M} M \frac{1}{L}\right]^{1/2} \tag{3.6.1}$$

But
$$\frac{LT^2}{M} = \frac{1}{E} \qquad (3.6.2)$$

Rewrite M as
$$M \sim \left[\frac{M}{L^3}L^3\right]$$

Then M can be written as
$$M = \rho l^3 \qquad (3.6.3)$$

Substitute (2) and (3) into (1)
$$T = \left[\frac{1}{E}\rho l^3 \frac{1}{l}\right]^{1/2} = l\sqrt{\frac{\rho}{E}}$$

Then the bending moment is
$$\bar{M} \sim ML^2T^{-2} = \frac{(\rho l^3)(l^2)}{l^2 \rho/E} = El^3$$

Then the first π term can be written as
$$\pi_1 = \frac{\bar{M}}{El^3}$$

Since $D \sim L \sim l$, $\pi_2 = \frac{D}{l}$ and $I \sim L^4 \sim l^4$, $\pi_3 = \frac{I}{l^4}$. From π_2
$$\frac{D_m}{l_m} = \frac{D_p}{l_p} \Rightarrow \frac{l_p}{l_m} = \frac{D_p}{D_m} = \frac{1.5D_m}{D_m} = 1.5$$

From π_1
$$\frac{\bar{M}_m}{E_m l_m^3} = \frac{\bar{M}_p}{E_p l_p^3}$$

Since $E_p = E_m$
$$\bar{M}_p = \bar{M}_m \frac{l_p^3}{l_m^3} = (3000)(1.5)^3 = 10,125 \; in-lb_f$$

3.7

```
      Parametr (n = 11, pi = 3.141592654)
      dimension phi (n,n)
      open(6,file='atila.out',status='new')
      r = 1.
      delx = 0.
      iter = 0
      eps = 0.001
c
c------------Boundary Conditions------------
c
      do 10 k = 1,n
```

```fortran
            phi(k,1) = 100. - 100. * delx
            phi(n,k) = 250. * delx
            phi(k,n) = 100. + 150. * delx
            phi(l,k) = 100.
            delx = delx + 0.1
      10    continue
c
c------------------Initial Conditions------------------
c
      do 20 i = 2, n-1
      do 20 j = 2, n-1
            phi(i,j) = 0.
      20    continue
c
c------------------Start Iteration--------------------
c
      50    iter = iter + 1
            flag = 1.
      do 30 i = 2, n-1
      do 30 j = 2, n-1
            temp = (phi(i, j + 1) + phi(i, j - 1) + r * (phi(i + 1, j)
     +      + phi(i - 1,j)))/ (2.* (1. + r))
            if(flag .eq. 0.) go to 35
            if(abs(temp - phi(i, j)) .gt. eps) flag = 0.
      35    phi(i, j) = temp
      30    continue
            if(iter .eq. 1.) then
              write(6, 140)
              write(6, 150)
              write(6, 130) ((phi(i, j), j = 1, n), i = 1, n)
            end if
c
            if(iter .eq. 500) go to 80
            if(flag .eq. 0.) go to 50
      80    write(6,90) iter
            write(6,100) eps
            write(6,110)
            write(6,120)
            write(6,130) ((phi(i,j),j=1,n),i=1,n)
      130   format(11(f6.1,x),/)
      90    format(//////,2x,'iter =',i4,/)
      100   format(2x,'eps =',f5.3,/)
      140   format(25x,'First Iteration')
      150   format(25x,'——————————',//)
      110   format(22x,'Temperature Distribution on the Plate')
      120   format(22x,'——————————————————',/)
            stop
            end
```

First iteration, eps=0.001

100.0	100.0	100.0	100.0	100.0	100.0	100.0	100.0	100.0	100.0	100.0
90.0	47.5	36.9	34.2	33.6	33.4	33.3	33.3	33.3	62.1	115.0
80.0	31.9	17.2	12.9	11.6	11.2	11.1	11.1	11.1	50.8	130.0
70.0	25.5	10.7	5.9	4.4	3.9	3.8	3.7	3.7	49.9	145.0
60.0	21.4	8.0	3.5	2.0	1.5	1.3	1.3	1.2	52.8	160.0
50.0	17.8	6.5	2.5	1.1	0.6	0.5	0.4	0.4	57.0	175.0
40.0	14.5	5.2	1.9	0.8	0.4	0.2	0.2	0.1	61.8	190.0
30.0	11.1	4.1	1.5	0.6	0.2	0.1	0.1	0.1	66.7	205.0
20.0	7.8	3.0	1.1	0.4	0.2	0.1	0.0	0.0	71.7	220.0
10.0	10.7	15.9	23.0	30.9	39.0	47.3	55.6	63.9	148.9	235.0
0.0	25.0	50.0	75.0	100.0	125.0	150.0	175.0	200.0	225.0	250.0

100th iteration, eps=0.001

100.0	100.0	100.0	100.0	100.0	100.0	100.0	100.0	100.0	100.0	100.0
90.0	92.5	95.0	97.5	100.0	102.5	105.0	107.5	110.0	112.5	115.0
80.0	85.0	90.0	95.0	100.0	105.0	110.0	115.0	120.0	125.0	130.0
70.0	77.5	85.0	92.5	100.0	107.5	115.0	122.5	130.0	137.5	145.0
60.0	70.0	80.0	90.0	100.0	110.0	120.0	130.0	140.0	150.0	160.0
50.0	62.5	75.0	87.5	100.0	112.5	125.0	137.5	150.0	162.5	175.0
40.0	55.0	70.0	85.0	100.0	115.0	130.0	145.0	160.0	175.0	190.0
30.0	47.5	65.0	82.5	100.0	117.5	135.0	152.5	170.0	187.5	205.0
20.0	40.0	60.0	80.0	100.0	120.0	140.0	160.0	180.0	200.0	220.0
10.0	32.5	55.0	77.5	100.0	122.5	145.0	167.5	190.0	212.5	235.0
0.0	25.0	50.0	75.0	100.0	125.0	150.0	175.0	200.0	225.0	250.0

3.8
```fortran
      parameter (n = 11, pi = 3.141592654)
      dimension phi (n,n)
      open(6,file='atila.out',status='new')
      r = 1.
      delx = 0.
      iter = 0
      eps = 0.001
c
c------------------Boundary Conditions----------------
c
      do 10 k = 1,n
      phi(k,1) = 0.
      phi(n,k) = 1000.* sin(pi * delx)
      phi(k,n) = 0.
      phi(1,k) = 0.
      delx = delx + 0.1
  10  continue
c
c------------------Initial Conditions----------------
c
      do 20 i = 2, n-1
      do 20 j = 2, n-1
      phi(i,j) = 0.
  20  continue
c
c------------------Start Iteration----------------
c
  50  iter = iter + 1
      flag = 1.
      do 30 i = 2, n-1
      do 30 j = 2, n-1
      temp = (phi(i, j + 1) + phi(i, j - 1) + r * (phi(i + 1, j)
     + phi(i - 1,j)))/ (2.* (1. + r))
      if(flag .eq. 0.) go to 35
      if(abs(temp - phi(i, j)) .gt. eps) flag = 0.
  35  phi(i, j) = temp
  30  continue
      if(iter .eq. 1.) then
        write(6, 140)
        write(6, 150)
        write(6, 130) ((phi(i, j), j = 1, n), i = 1, n)
      end if
c
      if(iter .eq. 500) go to 80
      if(flag .eq. 0.) go to 50
  80  write(6,90) iter
      write(6,100) eps
      write(6,110)
```

First iteration, eps=0.001

0.0	0.0	0.0	0.0	0.0	0.0	0.0	0.0	0.0	0.0	0.0
0.0	0.0	0.0	0.0	0.0	0.0	0.0	0.0	0.0	0.0	0.0
0.0	0.0	0.0	0.0	0.0	0.0	0.0	0.0	0.0	0.0	0.0
0.0	0.0	0.0	0.0	0.0	0.0	0.0	0.0	0.0	0.0	0.0
0.0	0.0	0.0	0.0	0.0	0.0	0.0	0.0	0.0	0.0	0.0
0.0	0.0	0.0	0.0	0.0	0.0	0.0	0.0	0.0	0.0	0.0
0.0	0.0	0.0	0.0	0.0	0.0	0.0	0.0	0.0	0.0	0.0
0.0	0.0	0.0	0.0	0.0	0.0	0.0	0.0	0.0	0.0	0.0
0.0	0.0	0.0	0.0	0.0	0.0	0.0	0.0	0.0	0.0	0.0
0.0	77.3	166.3	243.8	298.7	324.7	318.9	282.0	217.4	131.6	0.0
0.0	309.0	587.8	809.0	951.1	1000.0	951.1	809.0	587.8	309.0	0.0

107th iteration, eps=0.001

0.0	0.0	0.0	0.0	0.0	0.0	0.0	0.0	0.0	0.0	0.0
0.0	8.7	16.5	22.8	26.8	28.1	26.8	22.8	16.5	8.7	0.0
0.0	18.2	34.7	47.7	56.1	59.0	56.1	47.7	34.7	18.2	0.0
0.0	29.6	56.2	77.4	91.0	95.7	91.0	77.4	56.2	29.6	0.0
0.0	43.8	83.3	114.6	134.8	141.7	134.8	114.6	83.3	43.8	0.0
0.0	62.3	118.5	163.1	191.7	201.6	191.7	163.1	118.5	62.3	0.0
0.0	86.9	165.3	227.5	267.5	281.2	267.5	227.5	165.3	86.9	0.0
0.0	120.0	228.3	314.2	369.4	388.4	369.4	314.2	228.3	120.0	0.0
0.0	164.9	313.6	431.7	507.5	533.6	507.5	431.7	313.6	164.9	0.0
0.0	225.9	429.7	591.4	695.2	731.0	695.2	591.4	429.7	225.9	0.0
0.0	309.0	587.8	809.0	951.1	1000.0	951.1	809.0	587.8	309.0	0.0

```
      write(6,120)
      write(6,130) ((phi(i,j),j=1,n),i=1,n)
130   format(11(f6.1,x),/)
90    format(//////,2x,'iter =',i4,/)
100   format(2x,'eps =',f5.3,/)
110   format(22x,'Temperature Distribution on the Plate')
120   format(22x,'————————————————',/)
140   format(25x,'First Iteration')
150   format(25x,'——————————',//)
      stop
      end
```

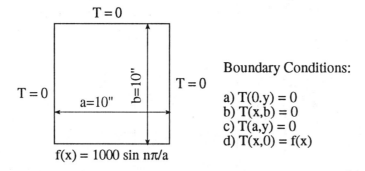

f(x) = 1000 sin nπ/a

Boundary Conditions:
a) T(0,y) = 0
b) T(x,b) = 0
c) T(a,y) = 0
d) T(x,0) = f(x)

Analitical Solution

Steady-state Heat conduction equation solution:

$$\frac{\partial^2 T}{\partial X^2} + \frac{\partial^2 T}{\partial Y^2} = 0 \qquad (3.8.1)$$

Assume solution $T(x,y) = X(x)Y(y)$

$$\frac{\partial^2 T}{\partial X^2} = X''Y$$

$$\frac{\partial^2 T}{\partial Y^2} = Y''X$$

Substituting into Eq. (3.8.1)

$$X''Y + Y''X = 0$$

or

$$\frac{X''}{X} = -\frac{Y''}{Y} = -\lambda$$

Solution of the above equation yields

$$T(x,y) = \sum_{n=1}^{\infty} \frac{1000}{\sinh \frac{n\pi}{a}b} \sin \frac{n\pi x}{a} \sinh \frac{n\pi}{a}(b-y)$$

For $n=1$, $a=1$, and $b=1$, we have the following form of solution

$$T(x,y) = \frac{1000}{\sinh \pi} \sin \pi x \sinh \pi(1-y)$$

The temperature at the center of the plate can be determined from the above equation as

$$T(0.5, 0.5) = \frac{1000}{\sinh \pi} \sin 0.5\pi \sinh \pi(0.5) = 199.2° F$$

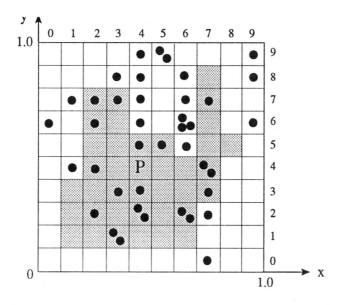

3.9

From Table 3.3, $n = 40$. Using the second digits of the respective random pair in Table 3.3 as the coordinates of the random points, it can easily be shown that $n' = 19$.

$$\text{Area} = \frac{n'}{n} = \frac{19}{40} = 0.475 \qquad (3.9.1)$$

0.4 Chapter 4

4.1

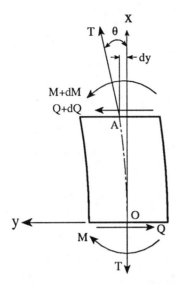

Taking the horizontal forces acting towards the left as positive, the first equilibrium equation can be formulated as:

$$\sum F_y = -Q + (Q + dQ) = 0 \tag{4.1.1}$$

from which it can easily be shown that

$$dQ = 0 \tag{4.1.2}$$

which implies that the shear force, Q, is constant. Letting the moments acting counter-clockwise to be positive, the second equilibrium equation can be formulated as:

$$\sum M_A = M + dM + Qdx - Tdy = 0 \tag{4.1.3}$$

Dividing the above equation by dx and rearranging yields

$$\frac{dM}{dx} - T\frac{dy}{dx} = -Q \tag{4.1.4}$$

Keeping in mind that Q is constant, taking the derivative of the above equation yields

$$\frac{d^2M}{dx^2} - T\frac{d^2y}{dx^2} = 0 \tag{4.1.5}$$

But the bending moment, M, is given by

$$M = EI\frac{d^2y}{dx^2} \tag{4.1.6}$$

therefore (4.1.5) can be rewritten as

$$EI\frac{d^4y}{dx^4} - T\frac{d^2y}{dx^2} = 0 \tag{4.1.7}$$

or on rearranging

$$y'''' - \frac{T}{EI}y'' = 0 \tag{4.1.8}$$

where the prime denotes a derivative with respect to x. Considering the T and EI as constant and defining $k^2 = T/EI$, equation (4.1.8) can be written as

$$y'''' - k^2 y'' = 0 \tag{4.1.9}$$

The solution of (9) can easily be shown to be

$$y(x) = C_1 + C_2 x + c_3 \cosh kx + C_4 \sinh kx \tag{4.1.10}$$

Assuming that the coordinate system is located at the ball joint, the following boundary conditions can be written:

$$x = 0: \quad y = 0; \quad \frac{dy}{dx} = 0; \quad \frac{d^2 y}{dx^2} = \frac{M}{EI} \tag{4.1.11}$$

$$x \to \infty: \quad \frac{dy}{dx} \cong \theta; \quad \frac{d^2 y}{dx^2} = 0 \Rightarrow M = 0 \tag{4.1.12}$$

The above boundary conditions can be used in (10) to yield

$$y(x) = -\frac{\theta}{k}[(1 - \cosh kx) - (kx - \sinh kx)] \tag{4.1.13}$$

Using (4.1.6) in (4.1.13) it can easily be shown that

$$M(x) = \frac{T\theta}{k}(\cosh kx - \sinh kx) \tag{4.1.14}$$

Now using (4.1.13) and (4.1.14) in (4.1.4) yields

$$Q(x) = EI\theta k^2 (\sinh kx - \cosh kx) \tag{4.1.15}$$

In summary, for the case of one point contact at the ball joint, the deflection, $y(x)$, bending moment, $M(x)$, and the shear force distribution, $Q(x)$, of a drill pipe are given by:

$$y(x) = -\frac{\theta}{k}[(1 - \cosh kx) - (kx - \sinh kx)] \tag{4.1.13}$$

$$M(x) = \frac{T\theta}{k}(\cosh kx - \sinh kx) \tag{4.1.14}$$

$$Q(x) = EI\theta k^2 (\sinh kx - \cosh kx) \tag{4.1.15}$$

4.2

Table 4.2.1 Fatigue Analysis based on SN-Curve

Angle θ	T=100kips	T=200kips
1	0.120	0.360×10^1
2	0.797×10^1	0.324×10^2
3	0.336×10^2	0.504×10^2
4	0.499×10^2	0.720×10^2

4.3
Note: If the cycles to failure, N, is greater than 10^6, then $n/N(\%)$ is set to zero. To make a genuine comparison of the above solution and Soln. 4.2, all cycles to failure, N, less than 1×10^4 should to be set to 1×10^4.

Table 4.3.1 Fatigue Analysis based on theory of fracture mechanics

Angle θ	T=100kips	T=200kips
1	0.345	0.124×10^1
2	0.535×10^1	0.157×10^2
3	0.221×10^2	0.894×10^2
4	0.708×10^2	0.596×10^3

4.4
STEEL A

$\sigma_{ys} = 1750\ MPa$, $K_{ic} = 87\ MPa\sqrt{m}$, $P = 8\ MPa$, $\rho = 7.7$, $E = 210$, $Cost = 450\ \$/Tonne$.

1st iteration

Assume deign stress, σ_D is

$$\sigma_D = 0.55\sigma_{ys} = 0.55 \times 1750 = 962.5\ MPa$$

Stress intensity equation

$$K_I = 1.1 M_K \sigma \sqrt{\pi \frac{a}{Q}}$$

can be modified to critical stress intensity equation by using the maximum design stress that the air vessel can safely withstand as

$$K_{IC} = 1.1 M_K \sigma_D \sqrt{\pi \frac{a}{Q}}$$

rearranging yields

$$\sigma_D = \frac{\sqrt{Q} K_{IC}}{1.1 M_K \sqrt{\pi a}}$$

where

$$\sqrt{Q} = \sqrt{\phi^2 - 0.212(\frac{\sigma_D}{\sigma_{ys}})^2}$$

and

$$\phi = \frac{3\pi}{8} + \frac{\pi}{8}(\frac{a}{c})^2$$

Since $\frac{a}{c}$ is assumed constant, ϕ is also constant

$$\phi = \frac{3\pi}{8} + \frac{\pi}{8}(0.5)^2 = 1.2763$$

$$Q = (1.2763)^2 - 0.212(0.55)^2 = 1.5648$$

$$\sqrt{Q} = 1.2509$$

For the first iteration, assume $M_K = 1.0$

$$\sigma_D = \frac{1.2509(87MPa\sqrt{m})}{(1.1)(1.0)\sqrt{\pi(0.008)}} = 624.1$$

$$\sigma_D = 624.1 MPa$$

From thin-walled pressure vessel theory,

$$\sigma_D = \frac{pD}{2t}$$

rearranging yields

$$t = \frac{pD}{2\sigma_D} = \frac{(14MPa)(1.0m)}{2 \times (624.1\ MPa)} = 0.0112$$

which yields

$$\frac{a}{t} = \frac{0.008}{0.0112} = 0.714$$

referring to Fig. 4.18, $M_K \approx 1.42$
and new ratio,

$$\frac{\sigma_D}{\sigma_{ys}} = \frac{624.1}{1750} = 0.36$$

Using these calculated values for M_K and $\frac{\sigma_D}{\sigma_{ys}}$

$$Q = (1.2763)^2 - 0.212(0.36)^2 = 1.6015$$

$$\sqrt{Q} = 1.2655$$

$$\sigma_D = \frac{(1.2655)(87)}{(1.1)(1.42)\sqrt{\pi(0.008)}} = 444.6$$

% error

$$\sigma_D = \frac{|(624.1 - 444.6)|}{624.1} \times 100 = 28.8\%$$

% error

$$M_K = \frac{|1.0 - 1.42|}{1.0} \times 100 = 42\%$$

For the next iteration, bisect the range of $\frac{\sigma_D}{\sigma_{ys}}$ as well as the range of M_K for initial guesses.

$$\frac{\sigma_D}{\sigma_{ys}} = \frac{(0.55 + 0.36)}{2} = 0.455$$

$$M_K = \frac{(1.0 + 1.42)}{2} = 1.21$$

Second iteration

Initial guesses are; $\frac{\sigma_D}{\sigma_y} = 0.455$ and $M_K = 1.21$

$$\phi = (1.2763)^2 - 0.212(0.455)^2 = 1.5851$$

$$\sqrt{\phi} = 1.2590$$

$$\sigma_D = \frac{(1.2590)(87)}{(1.1)(1.21)\sqrt{\pi(0.008)}} = 519.1\ MPa$$

$$t = \frac{(14)(1.0)}{2 \times 519.1} = 0.0135$$

$$\frac{a}{t} = \frac{0.008}{0.0135} = 0.593$$

Referring to Fig. 4.18, $M_K = 1.29$

$$\frac{\sigma_D}{\sigma_y} = \frac{519.1}{1750} = 0.297$$

$$Q = (1.2763)^2 - 0.212(0.297)^2 = 1.6102$$

$$\sqrt{Q} = 1.2689$$

$$\sigma_D = \frac{(1.2689)(87)}{(1.1)(1.29)\sqrt{\pi \times 0.008}} = 490.7\ MPa$$

% error

$$\sigma_D = \frac{|\ 519.1 - 490.7\ |}{519.1} 100 = 5.47\%$$

$$M_K = \frac{|\ 1.21 - 1.29\ |}{1.21} 100 = 6.61\%$$

bisect the range of $\frac{\sigma_D}{\sigma_{ys}}$ and M_K for the next iteration

$$\frac{\sigma_D}{\sigma_{ys}} = \frac{(0.455 + 0.297)}{2} = 0.376$$

$$M_K = \frac{(1.21 + 1.29)}{2} = 1.25$$

<u>Third iteration</u>
Initial guesses are: $\frac{\sigma_D}{\sigma_{ys}} = 0.376$ and $M_K = 1.25$

$$Q = (1.2763)^2 - 0.212(0.376)^2 = 1.5989$$

$$\sqrt{Q} = 1.2645$$

$$\sigma_D = \frac{(1.2645)(87)}{(1.1)(1.25)\sqrt{\pi \times 0.008}} = 504.7 \ MPa$$

$$t = \frac{(14)(1.0)}{2 \times 504.7} = 0.0139$$

$$\frac{a}{t} = \frac{0.008}{0.0139} = 0.576$$

Referring to Fig. 4.18, $M_K = 1.27$

$$\frac{\sigma_D}{\sigma_y} = \frac{504.7}{1750} = 0.2884$$

$$Q = (1.2763)^2 - 0.212(0.2884)^2 = 1.6113$$

$$\sqrt{Q} = 1.2694$$

$$\sigma_D = \frac{(1.2694)(87)}{(1.1)(1.27)(\sqrt{\pi \times 0.008})} = 498.7 MPa$$

% error
$$\sigma_D = \frac{|\ 504.7 - 498.7\ |}{504.7} \times 100 = 1.19\%$$

% error
$$M_K = \frac{|\ 1.25 - 1.27\ |}{1.25} \times 100 = 1.6\%$$

After convergence is achieved, the cost of constructing the pressure vessel out of sheet A, neglecting the end caps of the vessel, is estimated as

$$V = \frac{\pi}{4}(D_o^2 - D_i^2) \times L$$

where, $D_o = D_i + 2t$

$$D_o = 1.0 + 2(0.0139) = 1.028 m.$$

$$V = \frac{\pi}{4}[(1.028)^2 - (1)^2] \times 3 = 0.1335$$

$$m = \rho V = (7.7)(0.1335) = 0.6853 \ tonnes$$

$$Cost = (650)(0.6853) = \$668.0$$

For the other materials same calculations are performed and the results are shown in Table 4.4.1.

Table 4.4.1 Calculation of design parameters

Material	S_y	K_{ic}	Sigma	S_y/Sigma	Mk	Q	t	m	Cost
Steel # 1	1750	87	502.6	3.48	1.26	1.61	0.0139	1.0248	668.00
Steel # 2	1450	115	598.8	2.42	1.39	1.59	0.0177	0.8582	557.86
Steel # 3	1200	145	688.1	1.74	1.54	1.56	0.0102	0.7458	335.59
Steel # 4	1500	105	564.1	2.66	1.35	1.59	0.0124	0.9117	501.41
Steel # 5	1600	58	374.3	4.27	1.13	1.61	0.0187	1.3827	691.33
Steel # 6	1400	77	458.7	3.05	1.22	1.61	0.0153	1.1245	393.56

Table 4.4.2: Overall rating

Material	S_y/ρ		$\frac{K_{IC}}{S_y}\sqrt{mm}$		S_y/σ_d		$C(\rho/S_y)$		m (Kg)		Overall rating $G=\frac{3R_2+2(1-R_4)+4(1-R_5)}{12}$
	A_1	R_1	A_2	R_2	A_3	R_3	A_4	R_4	A_5	R_5	
Steel 3	156	0.80	3.9	1.76	1.74	0.57	2.88	1.00	0.7458	0.66	0.536
Steel 4	195	1.00	2.21	1.00	2.86	0.87	2.82	0.98	0.9117	0.81	0.534
Steel 6	182	0.93	1.74	0.79	3.05	1.00	1.923	0.67	1.1245	1.00	0.528

Steel #1, #2, and #5 are eliminated from the selection process because of their high cost. From the overall rating table, steel #3 may be the most suitable choice. It has low cost and low mass. Although its factor of safety is the lowest, it is still within the acceptable region for this application.

4.5

The mean stress can be evaluated as

$$\sigma_{mean} = \frac{T}{A} = \frac{100000}{\frac{\pi}{4}(5^2 - 4.276^2)} = 18959 psi = 18.96 kpsi \tag{4.5.1}$$

From which it follows:

$$\sigma_{max} = 22.6 + 18.96 = 41.56 kpsi \tag{4.5.2}$$

$$\Delta\sigma = 2\sigma_{av} = 160 kpsi \tag{4.5.3}$$

The thickness of the pipe is

$$t = \frac{5 - 4.226}{2} = 0.362 in \tag{4.5.4}$$

It is given that $a_i = 1/32 = 0.03125 in$

Material A:

$$\Phi = \frac{3\pi}{8} + \frac{\pi}{8}\left(\frac{a}{c}\right)^2 = \frac{3\pi}{8} + \frac{\pi}{8}(0.5)^2 = 1.276 \tag{4.5.5}$$

$$Q = \Phi^2 - 0.212\left(\frac{\sigma_{max}}{S_y}\right)^2 = (1.276)^2 - 0.212\left(\frac{41.56}{90}\right)^2 = 1.584 \tag{4.5.6}$$

Assuming $M_k = 1$

$$a_{cr} = \left(\frac{K_{IC}}{1.1 M_k \sigma_{max}}\right)^2 \frac{Q}{\pi} = \left(\frac{56}{1.1(1)41.56}\right)^2 \frac{1.584}{\pi} = 0.7564 in \tag{4.5.7}$$

Since $a_{cr} = 0.7564 > t = 0.362$, there will be a leak before failure.

$$\Delta a = \frac{0.362 - 0.03125}{2} = 0.1654 in \quad (4.5.8)$$

It then follows

$$a_{av} = a_i + \frac{\Delta a}{2} = 0.0312 + \frac{0.1654}{2} = 0.1139 in \quad (4.5.9)$$

$$\Delta K_I = 1.1 \Delta \sigma \left(\pi \frac{a_{av}}{Q} \right)^{0.5} = 1.1(45.2) \left(\pi \frac{0.1139}{1.584} \right)^{0.5} = 23.635 kpsi\sqrt{in} \quad (4.5.10)$$

It follows

$$\Delta N = \frac{\Delta a}{A(\Delta K_I)^m} = \frac{0.1654}{0.614 \times 10^{-10}(23.635)^{3.16}} = 123002 cycles \quad (4.5.11)$$

Now carrying out the second iteration:

$$a_i = 0.03125 + 0.1654 = 0.19665 in \quad (4.5.12)$$

$$a_{av} = 0.19665 + \frac{0.1654}{2} = 0.27935 in \quad (4.5.13)$$

$$\Delta K_I = 1.1(45.2) \left(\pi \frac{0.27935}{1.584} \right)^{0.5} = 37.00 kpsi\sqrt{in} \quad (4.5.14)$$

$$\Delta N = \frac{0.1654}{0.614 \times 10^{-10}(37.00)^{3.16}} = 29822 cycles \quad (4.5.15)$$

$$\sum N_A = 123002 + 29822 = 152884 cycles \quad (4.5.16)$$

Material B:

Following the same steps as for material A, it can easily be shown that:

$$\Phi = 1.276 \quad (4.5.17)$$

$$Q = 1.571 \quad (4.5.18)$$

Assuming $M_k = 1$

$$a_{cr} = 0.3455 in \quad (4.5.19)$$

Since $a_{cr} = 0.3455 < t = 0.362$, there will be a catastrophic failure before leaking.

$$\Delta a = 0.1571 in \quad (4.5.20)$$

It then follows

$$a_{av} = 0.1098 in \quad (4.5.21)$$

$$\Delta K_I = 23.30 kpsi\sqrt{in} \quad (4.5.22)$$

It follows

$$\Delta N = 122258 cycles \quad (4.5.23)$$

Now carrying out the second iteration:

$$a_i = 0.18835 in \quad (4.5.24)$$

$$a_{av} = 0.2669 in \quad (4.5.25)$$

$$\Delta K_I = 36.33 kpsi\sqrt{in} \quad (4.5.26)$$

$$\Delta N = 30046 cycles \tag{4.5.27}$$

$$\sum N_B = 152289 cycles \tag{4.5.28}$$

For unit length of material,

$$m = \rho V = 0.283\frac{\pi}{4}(5.0^2 - 4.276^2)(1) = 1.493 Ibm \tag{4.5.29}$$

Cost based on mass:
Material A: $(1.493)15 = \$2.24$
Material B: $(1.493)10 = \$1.49$
Cost criteria based on strength:
Material A: $1.5(0.283)/90 = 4.72 \times 10^{-3}$
Material B: $1.0(0.283)/80 = 3.54 \times 10^{-3}$
Optimum stress based on strength:
Material A: $90/0.283 = 318$
Material B: $80/0.283 = 282$
Safety factor analysis:
Material A: $90/41.56 = 2.16$
Material B: $90/41.56 = 1.92$
Overall rating:
Material A: 0.75
Material B: 0.59
Overall rating equation used:

$$G = \frac{1R_1 + 2R_2 + 3(1 - R_3) + 4R_4 + 2R_5}{12} \tag{4.5.30}$$

Table 4.5.1 Overall Rating

Material	Service Life $\sum N$		Strength S_y/ρ		Stress ρ/σ		Leak-Break a_{cr}/t		Safety S_y/σ_{max}		Overall Rating
	A_1	R_1	A_1	R_1	A_1	R_1	A_1	R_1	A_1	R_1	
A	152.9×10^3	1.0	318	1.0	4.7×10^{-3}	1.0	2.1	1.0	2.16	1.0	.75
B	152.3×10^{-3}	.99	282	.87	3.5×10^3	.74	.95	.45	1.92	.89	.59

Based on the analysis of materials A and B (cf. Table 4.5.1), the choice ofr the drill pipe material would be material A.

4.6
The stresses can be calculated using

$$\sigma = \frac{F}{(b - a_i)t} \tag{4.6.1}$$

From which it follows:

$$\sigma_{max} = \frac{F_{max}}{(b - a_i)t} = \frac{23040}{(2.4 - 0.3)0.12} = 91 \times 10^3 psi = 91 kpsi \tag{4.6.2}$$

$$\sigma_{min} = \frac{F_{min}}{(b - a_i)t} = \frac{17280}{(2.4 - 0.3)0.12} = 69 \times 10^3 psi = 69 kpsi \tag{4.6.3}$$

$$\sigma_{av} = \frac{\sigma_{max} - \sigma_{min}}{2} = \frac{91 - 69}{2} = 11 kpsi \tag{4.6.4}$$

It is given that $a_i = 0.3in$. Modify the stress-intensity equation to yield

$$K_{IC} = \sigma_{max}\sqrt{\pi a_{cr}} \tag{4.6.5}$$

$$a_{cr} = \frac{1}{\pi}\left(\frac{K_{IC}}{\sigma}\right)^2 = \frac{1}{\pi}\left(\frac{120}{91}\right)^2 = 0.55in \tag{4.6.6}$$

Assume $\Delta a = 0.02in$. It then follows

$$a_{av} = a_i + \frac{\Delta a}{2} = 0.3 + \frac{0.02}{2} = 0.3in \tag{4.6.7}$$

$$f(\frac{a_{av}}{b}) = 1.223 - 0.231\left(\frac{a_{av}}{2b}\right) + 10.550\left(\frac{a_{av}}{2b}\right)^2 - 21.710\left(\frac{a_{av}}{2b}\right)^3$$

$$+ 30.382\left(\frac{a_{av}}{2b}\right)^4 = 1.25 \tag{4.6.8}$$

From

$$K_I = \sigma\sqrt{\pi a}f(\frac{a_{av}}{b}) \tag{4.6.9}$$

it can be written

$$\Delta K_I = \Delta\sigma\sqrt{\pi a}f(\frac{a_{av}}{b}) = 22\sqrt{\pi 0.31}1.25 = 27.07 kpsi\sqrt{in} \tag{4.6.10}$$

From

$$\frac{da}{dN} = A(\Delta K_I)^m \tag{4.6.11}$$

it follows

$$\Delta N = \frac{\Delta a}{A(\Delta K_I)^m} = \frac{0.02}{0.66\times 10^{-8}(27.07)^{2.25}} = 1813 cycles \tag{4.6.12}$$

The above steps are repeated for increasing crack size, a_i, till the critical crack, a_{cr}, is reached. The iterations are shown in the following table.

Table 4.6.1 Iteration Table

Iteration	a_i	ΔK_I	ΔN	N
1	0.30	27.07	1813	1813
2	0.32	28.01	1679	3492
3	0.34	28.94	1560	5052
4	0.36	29.86	1454	6505
5	0.38	30.77	1359	7865
6	0.40	31.67	1274	9139
7	0.42	32.56	1197	10336
8	0.44	33.44	1127	11463
9	0.46	34.32	1063	12526
10	0.48	35.20	1004	13530
11	0.50	36.07	950	14480
12	0.52	36.94	901	15381
13	0.54	37.81	855	16236
14	0.55	38.25	833	17069

Since at $a_i = a_{cr} = 0.55in$, $N = 17069 cycles \leq 500000 cycles$, the design is not adequate for the prescribed life cycle. Recommendations for the change of design include among other: decreasing the fan rotation speed, increasing the fan tranverse dimension and increasing the K_{IC} value of the material.

4.7

From the thin-walled pressure vessel theory

$$\sigma = \frac{PD}{2t} \tag{4.7.1}$$

From which it follows:

$$\sigma_{max} = \frac{P_{max}D}{2t} = \frac{30 \times 1}{2 \times 1/16} = 240 kpsi \tag{4.7.2}$$

$$\sigma_{min} = \frac{P_{min}D}{2t} = \frac{10 \times 1}{2 \times 1/16} = 80 kpsi \tag{4.7.3}$$

$$\sigma_{av} = \frac{\sigma_{max} - \sigma_{min}}{2} = \frac{240 - 80}{2} = 80 kpsi \tag{4.7.4}$$

$$\Delta\sigma = 2\sigma_{av} = 160 kpsi \tag{4.7.5}$$

It is given that $a_i = 1/64 = 0.016 in$

$$\Phi = \frac{3\pi}{8} + \frac{\pi}{8}\left(\frac{a}{c}\right)^2 = \frac{3\pi}{8} + \frac{\pi}{8}(0.5)^2 = 1.276 \tag{4.7.6}$$

$$\sqrt{Q} = \left[\Phi^2 - 0.212\left(\frac{\sigma_{max}}{Sy}\right)^2\right]^{0.5} \tag{4.7.7}$$

$$Q = \Phi^2 - 0.212\left(\frac{\sigma_{max}}{Sy}\right)^2 = (1.276)^2 - 0.212\left(\frac{240}{255}\right)^2 = 1.44 \tag{4.7.8}$$

Assuming $M_k = 1$

$$a_{cr} = \left(\frac{K_{IC}}{1.1 M_k \sigma_{max}}\right)^2 \frac{Q}{\pi} = \left(\frac{79}{1.1(1)240}\right)^2 \frac{1.44}{\pi} = 0.041 \tag{4.7.9}$$

Assume $\Delta a = 0.005 in$. It then follows

$$a_{av} = a_i + \frac{\Delta a}{2} = 0.016 + \frac{0.005}{2} = 0.0185 in \tag{4.7.10}$$

$$\Delta K_I = 1.1\Delta\sigma\left(\pi\frac{a_{av}}{Q}\right)^{0.5} = 1.1(160)\left(\pi\frac{0.0185}{1.44}\right)^{0.5} = 35.36 kpsi\sqrt{in} \tag{4.7.11}$$

From

$$\frac{da}{dN} = A(\Delta K_I)^m \tag{4.7.12}$$

it follows

$$\Delta N = \frac{\Delta a}{A(\Delta K_I)^m} = \frac{0.005}{0.66 \times 10^{-8}(35.36)^{2.25}} = 248 cycles \tag{4.7.13}$$

The above steps are repeated for increasing crack size, a_i, till the critical crack, a_{cr}, is reached. The iterations are shown in the following table.

Table 4.7.1 Iteration Table

Iteration	a_i	ΔK_I	ΔN	N
1	0.016	35.36	248	248
2	0.021	39.85	190	438
3	0.026	43.89	153	591
4	0.031	47.58	127	718
5	0.036	51.01	107	827
6	0.041	54.22	95	922

The number of cycles before failure is $N = 922\ cycles$.

4.8

For the given material properties, it can easily be shown that the elements of the reduced stiffness matrix are:

$$Q_{11} = \frac{20 \times 10^6}{1 - 0.29(0.02176)} = 20.13 \times 10^6 \tag{4.8.1}$$

$$Q_{22} = \frac{1.5 \times 10^6}{1 - 0.29(0.02176)} = 1.51 \times 10^6 \tag{4.8.2}$$

$$Q_{12} = \frac{1.5 \times 10^6 (0.29)}{1 - 0.29(0.02176)} = 0.44 \times 10^6 \tag{4.8.3}$$

$$Q_{66} = 1.0 \times 10^6 \tag{4.8.4}$$

For the given material properties, it can further be shown that the transformed elements of the reduced stiffness matrix are:

$$\bar{Q}_{11} = 12.33 \times 10^6; \quad \bar{Q}_{12} = 3.58 \times 10^6; \quad \bar{Q}_{22} = 3.02 \times 10^6 \tag{4.8.5}$$

$$\bar{Q}_{16} = 5.85 \times 10^6; \quad \bar{Q}_{26} = 2.22 \times 10^6; \quad \bar{Q}_{66} = 4.14 \times 10^6 \tag{4.8.6}$$

The along the principal material axes can be formulated as

$$\begin{pmatrix} \sigma_1 \\ \sigma_2 \\ \tau_{12} \end{pmatrix} = [T] \begin{pmatrix} \sigma_x \\ \sigma_y \\ \tau_{xy} \end{pmatrix} \tag{4.8.7}$$

$$\begin{pmatrix} \sigma_1 \\ \sigma_2 \\ \tau_{12} \end{pmatrix} = \begin{pmatrix} \bar{Q}_{11} & \bar{Q}_{12} & \bar{Q}_{16} \\ \bar{Q}_{12} & \bar{Q}_{22} & \bar{Q}_{26} \\ \bar{Q}_{16} & \bar{Q}_{26} & \bar{Q}_{66} \end{pmatrix} \begin{pmatrix} m^2 & n^2 & 2mn \\ n^2 & m^2 & -2mn \\ -mn & mn & (m^2 - n^2) \end{pmatrix} \begin{pmatrix} \varepsilon_x \\ \varepsilon_y \\ \gamma_{xy} \end{pmatrix} \tag{4.8.8}$$

where $n = \cos 45°$ and $n = \sin 45°$. The above yields

$$\begin{pmatrix} \sigma_1 \\ \sigma_2 \\ \tau_{12} \end{pmatrix} = \begin{pmatrix} 52.47 \times 10^2 \\ 3.34 \times 10^2 \\ -1.92 \times 10^2 \end{pmatrix} psi \tag{4.8.9}$$

4.9

For the given material properties, it can further be shown that the transformed elements of the compliance matrix are:

$$\bar{S}_{11} = 0.4219 \times 10^{-6}; \quad \bar{S}_{12} = -0.0780 \times 10^{-6}; \quad \bar{S}_{22} = 0.4219 \times 10^{-6} \tag{4.9.1}$$

$$\bar{S}_{16} = -0.3083 \times 10^{-6}; \quad \bar{S}_{26} = -0.3083 \times 10^{-6}; \quad \bar{S}_{66} = 0.7747 \times 10^{-6} \tag{4.9.2}$$

The strains in the laminar are

$$\begin{pmatrix} \varepsilon_x \\ \varepsilon_y \\ \gamma_{xy} \end{pmatrix} = \begin{pmatrix} \bar{S}_{11} & \bar{S}_{12} & \bar{S}_{16} \\ \bar{S}_{12} & \bar{S}_{22} & \bar{S}_{26} \\ \bar{S}_{16} & \bar{S}_{26} & \bar{S}_{66} \end{pmatrix} \begin{pmatrix} \sigma_x \\ \sigma_y \\ \tau_{xy} \end{pmatrix} \tag{4.9.3}$$

The above yields

$$\begin{pmatrix} \varepsilon_x \\ \varepsilon_y \\ \gamma_{xy} \end{pmatrix} = \begin{pmatrix} 26.533 \times 10^{-4} \\ -3.787 \times 10^{-4} \\ -20.391 \times 10^{-4} \end{pmatrix} psi \tag{4.9.4}$$

4.10

The stresses in the principal material axes of symmetry are

$$\begin{pmatrix} \sigma_1 \\ \sigma_2 \\ \tau_{12} \end{pmatrix} = \begin{pmatrix} m^2 & n^2 & 2mn \\ n^2 & m^2 & -2mn \\ -mn & mn & (m^2 - n^2) \end{pmatrix} \begin{pmatrix} \sigma_0 \\ 0 \\ 0 \end{pmatrix} \tag{4.10.1}$$

where $m = \cos 30° = \sqrt{3}/2$ and $m = \sin 30° = 1/2$

$$\begin{pmatrix} \sigma_1 \\ \sigma_2 \\ \tau_{12} \end{pmatrix} = \begin{pmatrix} (\sqrt{3}/2)^2 \\ (1/2)^2 \\ -(1/2)(\sqrt{3}/2) \end{pmatrix} \sigma_0 \tag{4.10.2}$$

(a)
$$(\sqrt{3}/2)^2 \sigma_0 \geq \sigma_1^2 = 200 kpsi \tag{4.10.3}$$

$$\sigma_0 \geq (2/\sqrt{3})^2 (200) = \frac{800}{3} kpsi \tag{4.10.4}$$

(b)
$$(1/2)^2 \sigma_0 \geq \sigma_2^2 = 100 kpsi \tag{4.10.5}$$

$$\sigma_0 \geq 2^2 (100) = 400 kpsi \tag{4.10.6}$$

(c)
$$-(1/2)(\sqrt{3}/2)\sigma_0 \leq \tau_{12}^* = -20 kpsi \tag{4.10.7}$$

$$\sigma_0 \leq 2(2/\sqrt{3})20 = 46.19 kpsi \tag{4.10.8}$$

The above three cases, the smallest σ_0 at which the laminate will fail is $\sigma_0 = 46.19 kpsi$.

4.11

This is an open ended problem. Generally, the selection of a suitable material should be done using an "Overall Rating Table." In this table, the three picked materials should be compared on the basis of design requirements and material selection factors. For this given problem, since a flactuating force has been specified, it is essential to inlcude the fatigue strength in the material design requirements. For example, one could select a A514-B steel of A533-B steel [Rolf and Barson (1977), *Fracture and Fatigue Control in Structures*, pp. 237–238].

0.5 Chapter 5

5.1.a

Table 5.1.1 Vehicle LCC W/O Financing Costs

End of Year A	O & M Cost At End of Year B	Cumulative O & M Cost $C = \sum B$	Avg O & M Cost $D = C/A$	Capital Value $E = 0.85 E_{n-1}$	Capital Cost $F = \$6000 - E$	Avg Capital Cost $G = F/A$	Avg LCC $H = D + G$
1	$ 1300	$ 1300	$ 1300	$ 5100	$ 900	$ 900	$ 2200
2	1400	2700	1350	4335	1665	832.50	2182.50
3	1500	4200	1400	3684.75	2315.85	771.75	2171.75
4	1600	5800	1450	3132.04	2867.96	717	2167
5	1700	7500	1500	2662.23	3337.77	667.55	2167.55
6	1800	9300	1550	2262.90	3737.10	622.85	2172.85

a1) Minimum Cost Life = 4 Years
   Minimum Annual LCC = $ 2167

Table 5.1.2 Vehicle LCC W/O Financing Costs

End of Year A	O & M Cost At End of Year B	Cumulative O & M Cost $C = \sum B$	Avg O & M Cost $D = C/A$	Capital Value $E = 0.82 E_{n-1}$	Capital Cost $F = \$10000 - E$	Avg Capital Cost $G = F/A$	Avg LCC $H = D + G$
1	$ 1200	$ 1200	$ 1200	$ 8200	$ 1800	$ 1800	$ 3000
2	1400	2600	1300	6724	3276	1638	2938
3	1600	4200	1400	5513.68	4486.32	1495.44	2895.44
4	1800	6000	1500	4521.22	5478.78	1369.70	2869.70
5	2000	8000	1600	3707.40	6292.60	1258.52	2858.52
6	2200	10200	1700	3040.07	6959.93	1159.99	2859.99

a2) Minimum Cost Life = 5 Years
   Minimum LCC = $ 2858.52

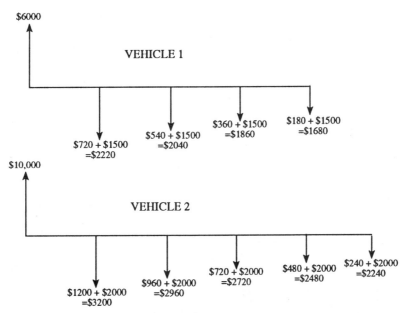

Note: Vehicle could also be paid off in equal monthly payments of $199.28 and $332.14, respectively.

5.1.b

For part b of this problem some assumptions must be made. Assume both vehicles will be paid-off over 3 years. Until vehicles are paid-off they belong to the financing organization.

Table 5.1.3 Vehicle LCC Including Financing

End of Year A	O & M Cost At End of Year B	Cumulative O & M Cost $C = \sum B$	Avg O & M Cost $D = C/A$	Total Cumulative Payment $E = \sum_1^n (P+i)$	Salvage Value F	Capital Cost $G = E - F$	Avg Capital Cost $H = G/A$	Avg LCC $I = D + H$
1	$ 1300	$ 1300	$ 1300	$ 2391.41	-	$ 2391.41	$ 2391.41	$ 3691.41
2	1400	2700	1350	4782.82	-	4782.82	2391.41	3741.41
3	1500	4200	1400	7174.23	3684.75	3489.48	1163.16	2563.16
4	1600	5800	1450	7174.23	3132.04	4042.19	1010.55	2460.55
5	1700	7500	1500	7174.23	2662.23	4512.00	902.40	2402.40
6	1800	9300	1550	7174.23	2262.90	4911.33	818.56	2368.56
7	1900	11200	1600	7174.23	1923.47	5250.77	750.11	2350.11
8	2000	13200	1650	7174.23	1634.95	5539.28	692.41	2342.41
9	2100	15300	1700	7174.23	1389.71	5784.52	642.72	2342.72

b1) Minimum Cost Life = 8 Years
    Minimum Annual LCC = $ 2342.41

Table 5.1.4 Vehicle LCC Including Financing

End of Year A	O & M Cost At End of Year B	Cumulative O & M Cost $C = \sum B$	Avg O & M Cost $D = C/A$	Total Cumulative Payment $E = \sum_1^n (P+i)$	Salvage Value F	Capital Cost $G = E - F$	Avg Capital Cost $H = G/A$	Avg LCC $I = D + H$
1	$ 1200	$ 1200	$ 1200	$ 3985.68	-	$ 3985.68	$ 3985.68	$ 5185.68
2	1400	2600	1300	7971.36	-	7971.36	3985.68	5285.68
3	1600	4200	1400	11957.04	5513.68	6443.36	2147.79	3547.79
4	1800	6000	1500	11957.04	4521.22	7435.82	1858.96	3358.96
5	2000	8000	1600	11957.04	3707.40	8249.64	1649.93	3249.93
6	2200	10200	1700	11957.04	3040.07	8916.97	1486.16	3186.16
7	2400	12600	1800	11957.04	2492.86	9464.18	1352.03	3152.03
8	2600	15200	1900	11957.04	2044.15	9912.89	1239.11	3139.11
9	2800	18000	2000	11957.04	1676.20	10280.83	1142.31	3142.31

b2) Minimum Cost Life = 8 Years
    Minimum LCC = $ 3139.11

Based on this cost analysis the $6000 vehicle would be the best buy; however, comfort and prestige might justify buying the more expensive vehicle depending on the buyer's priorities.

5.2

$$\frac{20,000}{12} = 1667 gal/yr$$

Annual savings = $0.5(1667) = $833.33

10 yr savings = $8333

$$P = A[\frac{(1+i)^n - 1}{i(1+i)^n}]$$
$$= 833.33[\frac{(1.12)^{10} - 1}{0.12(1.12)^{10}}]$$
$$= \$4708.50$$

Present value of salvage

$$P = F/(1+i)^n$$

$$= 0.15(\text{Conv. Cost})[\frac{1}{(1+0.12)^{10}}]$$
$$= 0.0483 \text{Conv. Cost}$$
$$\text{Conv. Cost} = \$4708.50 + 0.0483 \text{Conv. Cost}$$
$$= \$4947.44$$

Note:
This solution assumes that the time value of money also applies to the decreasing salvage value of the system.

5.3

$$P = A[\frac{(1+i)^n - 1}{i(1+i)^n}] + \frac{F}{(1+i)^n}$$
$$= (9600 - 2000)[\frac{(1+0.15)^{10} - 1}{0.15(1+0.15)^{10}}] + \frac{70,000}{(1+0.15)^{10}}$$
$$= 38,142.64 + 17,302.93$$
$$= \$55,445.57$$

5.4

Assume $i = 15\%$

$$\$15,000 + \$8000(0.8696) \neq \$10,000(0.7561) + \$12,000(0.6576) + \$15,000(0.5718) + \$15,000(0.4972)$$

$$\$21,957 \neq \$31,487$$

Assume $i = 20\%$

$$\$15,000 + \$8000(0.8333) \neq 10,000(0.6944) + \$12,000(0.5787) + \$15,000(0.4823) + \$15,000(0.4019)$$

$$\$21,666 \neq \$27,151$$

Assume $i = 28\%$

$$\$15,000 + \$*000(0.7813) \neq 10,000(0.6104) + \$12,000(0.4768) + \$15,000(0.3725) + \$15,000(0.2710)$$

$$\$21,250 \neq \$21,478$$

Assume $i = 30\%$

$$\$15,000 + \$8000(0.7692) = \$10,000(0.5917) + \$12,000(0.4552) + \$15,000(0.3501) + \$15,000(0.2693)$$

$$21,154 \neq \$20,670$$

$$i = 28 + 2[\frac{(21,478 - 21,250) - 0}{(21,478 - 21,250) - (20,670 - 21,154)}]$$
$$= 28 + 2(0.32)$$
$$= 28.64\%$$

5.5

A.

$$-\$1,000 + \$500(0.8696) + \$300(0.7561) + 200(0.6575) + 200(0.5718) + 200(0.4972)$$

$$\$6.92 \geq 0$$

Therefore, n = 5 years.

B.

$$-\$2,000 + 600(0.8696) + \$500(0.7561) + \$900(0.6575) + \$1,000(0.5718)$$

$$\$63.36 \geq 0$$

Therefore, n = 4 years.

C.

$$-\$900 - \$300(0.8696) + \$600(0.7561) + \$600(6575)$$

$$-\$311.42 \neq 0$$

Therefore, $n = \infty$

5.6

$$\text{Purchasing} = A$$
$$= P[\frac{(1+i)^n(i)}{(1+i)^n - 1}]$$
$$= \$40,000[\frac{0.12(1.12)^{10}}{(1.12)^{10} - 1}]$$
$$= \$7,079.37$$

Present worth of income tax,

$$P = \sum_{1}^{10}[ColG(1+0.12)^{-n}]$$
$$= \$3623.81(0.8929) + \$3623.81(0.7972) + \$3623.81(0.7118) +$$
$$\quad \$3623.81(0.6355) + \$3623.81(0.5674) + \$3623.81(0.5066) +$$
$$\quad \$3623.81(0.4523) + \$3623.81(0.4039) + \$3623.81(0.3606) +$$
$$\quad \$3623.81(0.3220)$$

Year End A	Payment B	Maintenance C	Depreciation D	Taxable Income $E = C + D + B$	Income Tax Rate F	Income Tax $G = F \times E$
1	$ 7079.37	$ 1000	$ 4000	$ 12,079.37	0.30	$ 3623.81
2	$ 7079.37	$ 1000	$ 4000	$ 12,079.37	0.30	$ 3623.81
3	$ 7079.37	$ 1000	$ 4000	$ 12,079.37	0.30	$ 3623.81
4	$ 7079.37	$ 1000	$ 4000	$ 12,079.37	0.30	$ 3623.81
5	$ 7079.37	$ 1000	$ 4000	$ 12,079.37	0.30	$ 3623.81
6	$ 7079.37	$ 1000	$ 4000	$ 12,079.37	0.30	$ 3623.81
7	$ 7079.37	$ 1000	$ 4000	$ 12,079.37	0.30	$ 3623.81
8	$ 7079.37	$ 1000	$ 4000	$ 12,079.37	0.30	$ 3623.81
9	$ 7079.37	$ 1000	$ 4000	$ 12,079.37	0.30	$ 3623.81
10	$ 7079.37	$ 1000	$ 4000	$ 12,079.37	0.30	$ 3623.81

$P = \$20,475$ This is the amount saved in income taxes.
Total cost $= \$40,000 - 20,475 = \$19,524.75$
Total cost to rent before tax,

$$A[\frac{(1+i)^n - 1}{i(1+i)^n}] = 1000[\frac{(1.01)^{120} - 1}{0.01(1.01)^{120}}] = \$69,700$$

If the rental expense can be deducted from income tax,
Deduction $= \$12,000(0.3)(0.8929) \cdots \$3600(0.3220) = \$20,340$
Total cost to rent $\$69,700 - \$20,340 = \$49,359$
Total cost to buy $= \$19,524$
Therefore, best to buy.

5.7

Cost of leasing $= 36(375) = \$ 13,500$
Cost of buying :

$$\begin{aligned} A &= P[\frac{i(1+i)^n}{(1+i)^n - 1}] = \$17,100[\frac{\frac{0.10}{12}(1 + \frac{0.10}{12})^{36}}{(1 + \frac{0.10}{12})^{36} - 1}] \\ &= \$17,100(0.03227) \\ &= \$551.77/\text{month} \end{aligned}$$

$$\begin{aligned} \text{Total payments} &= 36(551.77) \\ &= \$19,864 \\ \text{Total cost} &= \$19,864 + 0.05(18,000) - 8000 \\ &= \$12,764 \end{aligned}$$

Therefore, cheaper to buy.
Note : These calculations do not take into account the value of money (%) to the buyer nor do they account for cost of living (COL) factors.

5.8

A1)
Assume $i = 15 \%$

$$\$2000 \neq \$750(0.8696) + \$750(0.7561) + \$750(0.6575) + \$750(0.5718)$$
$$\$2000 \neq \$2141$$

Assume $i = 20\%$

$$\$2000 = \$750(0.8333) + \$750(0.6944) + \$750(0.5787) + \$750(0.4823)$$
$$\$2000 \neq \$1941$$
$$i = 15 + 5[\frac{(2141 - 2000) - 0}{(2141 - 2000) - (1941 - 2000)}]$$
$$= 15 + 5(0.705)$$
$$= 18.53\%$$

A2)
Assume $i = 20\%$

$$\$2500 \neq \$500(0.833) + \$1200(0.6944) + \$1200(0.5787) + \$1200(0.4823)$$
$$\$2500 \neq \$2523$$

Assume $i = 25\%$

$$\$2500 \neq \$500(0.8) + \$1200(0.64) + \$1200(0.512) + \$1200(0.4096)$$
$$\$2500 \neq \$2274$$

$$i = 20 + 5[\frac{(2523 - 2500) - 0}{(2523 - 2500) - (2274 - 2500)}]$$
$$= 20 + 5(0.09237)$$
$$= 20.46\%$$

A2 - A1)
Assume $i = 10\%$

$$\$500 = -\$250(0.9091) + \$450(0.8264) + \$450(0.7513) + \$450(0.6830)$$
$$\$500 \neq \$790$$

Assume $i = 15\%$

$$\$500 \neq -\$250(0.8696) + \$450(0.7561) + \$450(0.6575) + \$450(0.5718)$$
$$\$500 \neq \$676$$

Assume $i = 20\%$

$$\$500 \neq -\$250(0.833) + \$450(0.6944) + \$450(0.5787) + \$450(0.4823)$$
$$\$500 \neq \$581$$

Assume $i = 25\%$

$$\$500 \neq -\$250(0.8) + \$450(0.64) + \$450(0.512) + \$450(0.4096)$$
$$\$500 \neq \$503$$

Assume $i = 26\%$

$$\$500 = -\$250(0.7937) + \$450(0.6299) + \$450(0.5) + \$450(0.3968)$$
$$\$500 \neq \$489$$
$$i = 25 + 1[\frac{3-0}{3+11}]$$
$$= 25.21\%$$

5.9

Alt 1 :

$$P/A = [\frac{(1+i)^n - 1}{i(1+i)^n}] = \frac{\$10,000}{\$1424} = 7.0225$$
$$= 7.0225i(1+i)^{15} = (1+i)^{15} - 1$$

Assume $i = 15\%$, then $7.518 \neq 7.137$
Assume $i = 10\%$, then $2.9335 \neq 3.1772$

$$i = 10 + 5[\frac{(3.1772 - 2.9335) - 0}{(3.1772 - 2.9335) - (7.137 - 7.518)}]$$
$$= 11.95\%$$

For Alt 1, $i < 12\%$ and can be eliminated from further consideration.

Alt 2 :

$$P/A = \frac{\$6000}{\$1026} = 5.848$$
$$= 5.848i(1+i)^{15} = (1+i)^{15} - 1$$

Assume $i = 15\%$, then $7.137 \neq 7.137$
Therefore i for Alt 2 is 15%

Alt 3 :

$$P/A = \frac{4000}{742} = 5.391$$
$$= 5.391i(1+i)^{15} = (1+i)^{15} - 1$$

Assume $i = 20\%$, then $16.612 \neq 14.407$
Assume $i = 15\%$, then $6.580 \neq 7.137$

$$i = 15 + 5[\frac{(7.137 - 6.58) - 0}{(7.137 - 6.58) - (14.407 - 16.612)}]$$
$$= 16.01\%$$

Alt 4

$$P/A = \frac{2000}{356} = 5.618$$
$$= 5.618i(1+i)^{15} = (1+i)^{15} - 1$$

	5	4	3	2
Cost	$ 1,000	$ 2,000	$ 4000	$ 6000
Uniform Annual Benefits	$ 256	$ 356	$ 742	$ 1,026
Rate of Return	24.38 %	15.82 %	16.01 %	15 %

	Increment 4 - 5	Increment 3 - 5	Increment 2 - 3
Δ Cost	$ 1,000	$ 3,000	$ 2,000
Δ Annual Benefit	$ 100	$ 486	$ 284
Δ Rate of Return	5.19 %	13.0 %	10.69 %

Assume $i = 15\%$, then $6.857 \neq 7.137$
Assume $i = 16\%$, then $8.3286 \neq 8.266$

$$i = 15 + 1[\frac{0.28 - 0}{0.28 + 0.0626}]$$
$$= 15.82\%$$

Alt 5

$$P/A = \frac{1000}{256} = 3.90625$$
$$= 3.90625i(1+i)^{15} = (1+i)^{15} - 1$$

Assume $i = 20\%$, then $12.037 \neq 14.407$
Assume $i = 25\%$, then $27.756 \neq 27.4217$

$$i = 20 + 5[\frac{2.37 - 0}{2.37 + 0.3343}]$$
$$= 24.38\%$$

Increment 4 - 5

$$P/A = \frac{\$1,000}{\$100} = 10$$
$$= 10i(1+i)^{15} = (1+i)^{15} - 1$$

Assume $i = 10\%$, then $4.177 \neq 3.177$
Assume $i = 5\%$, then $1.039 \neq 1.079$

$$i = 5 + 5[\frac{0.04 - 0}{0.04 + 1}] = 5.19\%$$

Increment 3 - 5

$$P/A = \frac{\$3,000}{\$486} = 6.173$$
$$= 6.173i(1+i)^{15} = (1+i)^{15} - 1$$

Assume $i = 15\%$, then $7.535 \neq 7.137$
Assume $i = 10\%$, then $2.579 \neq 3.177$

$$i = 10 + 5\left[\frac{0.5984 - 0}{0.5984}\right] = 13.0\%$$

Increment 2 - 3

$$P/A = 7.042$$
$$7.042i(1+i)^{15} = (1+i)^{15} - 1$$

Assume $i = 10\%$ $2.942 \neq 3.177$
Assume $i = 15\%$ $8.595 \neq 7.137$

$$i = 10 + 5\left[\frac{0.235 - 0}{0.235 + 1.458}\right] = 10.69\%$$

Note :
Since increment 4 - 5 has a ROR < MARR Alternative 4 is discarded. Also, Increment 2 - 3 has a ROR < MARR. Thus, Alternative 2 is discarded and Alternative 3 is selected as the best investment.

5.10

$$\begin{aligned}
TTU : \text{Expected value} &= \text{Outcome if TTU wins} \times P(\text{TTU winning}) \\
&+ \text{Outcome if TTU loses} \times P(\text{TTU losing}) \\
&= 0.25(25.00) + 0.75(0.00) = \$6.25 \\
UT : \text{Expected value} &= 0.2(30.00) + 0.8(0.00) \\
&= \$15.00 \\
UH : \text{Expected value} &= 0.15(\triangle 0.00) + 0.85(0.00) \\
&= \$6.00
\end{aligned}$$

5.11

$$PW_{(net1)} = -\$1200 + \$267 \left(\begin{array}{c} P/A, 15, 10 \\ 5.0188 \end{array}\right) = \$140.01$$

$$\frac{PW_{net1}}{PW_{cost1}} = \frac{140.01}{1200} = 11.7\%$$

$$PW_{net2} = -\$2000 + \$521.20 \left(\begin{array}{c} P/A, 15, 8 \\ 4.4873 \end{array}\right) = \$338.79$$

$$\frac{PW_{net2}}{PW_{cost2}} = \frac{338.79}{2000} = 16.9\%$$

$$PW_{(net3)} = -\$500 + \$185.90 \left(\begin{array}{c} P/A, 15, 5 \\ 3.3522 \end{array}\right) = \$123.17$$

$$\frac{PW_{net3}}{PW_{cost3}} = \frac{123.17}{500} = 24.6\%$$

$$PW_{net4} = -\$2000 + \$477 \left(\begin{array}{c} P/A, 15, 10 \\ 5.0188 \end{array}\right) = \$393.97$$

$$\frac{PW_{net4}}{PW_{cost4}} = \frac{393.97}{2000} = 19.7\%$$

$$PW_{net5} = -\$2000 + \$438.5 \left(\begin{array}{c} P/A, 15, 10 \\ 5.0188 \end{array}\right) + \$1000 \left(\begin{array}{c} P/F, 15, 10 \\ 0.2472 \end{array}\right)$$

$$= \$447.93$$

$$\frac{PW_{net5}}{PW_{cost5}} = \frac{447.83}{2000} = 22.4\%$$

$$PW_{net6} = -\$1,500 + 225 \left(\begin{array}{c} P/A, 15, 6 \\ 3.7845 \end{array}\right) + 1500 \left(\begin{array}{c} P/F, 15, 6 \\ 0.4323 \end{array}\right) = 0$$

Project ranking
3
5
4
2
1
6

Fund projects 3, 5, 4 & 2 in that order.

5.12
a. Assume ROR = 10%

$$-\$50,000 + \$7000 \left(\begin{array}{c} P/F, 10, 15 \\ 0.2394 \end{array}\right) + (\$8000 - \$2700) \left(\begin{array}{c} P/A, 10, 15 \\ 7.6061 \end{array}\right) \neq 0$$

$$-8011 \neq 0$$

Assume ROR = 5%

$$-\$50,000 + \$7000 \left(\begin{array}{c} P/F, 5, 15 \\ 0.4810 \end{array}\right) + \$5300 \left(\begin{array}{c} P/A, 5, 15 \\ 10.3797 \end{array}\right) \neq 0$$

$$+8379 \neq 0$$

$$i^* = 5 + 5\left[\frac{8379 - 0}{8011 + 8379}\right] = 7.56\%$$

Not a good investment based on ROR

b.
$$PW_{7.56\%} = -\$50,000 + \$7000 \left(\begin{array}{c} P/F, 7.56, 15 \\ 0.33515 \end{array}\right) + \$5300 \left(\begin{array}{c} P/A, 7.56, 15 \\ 8.7943 \end{array}\right)$$

$$= -\$1044$$

Not a good investment based on PW.

c.
$$A = -\$50,000 \begin{pmatrix} A/P, 7.56, 15 \\ 0.11370 \end{pmatrix} + \$7000 \begin{pmatrix} A/F, 7.56, 15 \\ 0.03811 \end{pmatrix} + \$5300$$
$$= -\$118.73$$

Not a good investment based on annual worth.

d.
$$F = \$7000 + \%5300 \begin{pmatrix} F/A, 7.56, 15 \\ 26.24 \end{pmatrix} - \$50,000 \begin{pmatrix} F/P, 7.56, 15 \\ 2.98375 \end{pmatrix}$$
$$= -\$3,115$$

Not a good investment based on future worth.

e.
$$\frac{PW_{net}}{PW_{cost}} = \frac{-\$1044}{-\$50,000} = 2.1\%$$

Not a good investment based on $\frac{PW_{net}}{PW_{cost}}$.

5.13
$Q_T = 429{,}748$ Btu/container
Energy required for:
  Cooling tower fan = 0.25 hp
  Pump 1, 2, 3 = 3 × 0.04 = 0.12 hp
  Blower = 7.5 Hp
The total Hp is 0.25 + 0.12 + 7.5 = 7.87 hp. Or
W_T (7.87)(0.746)(38) = 223.10 kWh, or 761440.3 Btu.
$W_T + Q_T = 429{,}748 + 761440.3 = 1{,}191{,}188.3$ Btu/run.

The drying operation cost, DC_{new}, is
$DC_{new} = (223.10)(0.04) + (0.429748)(3.48) = \$10.574$/run

Annual operation cost, AC_{new}:
$AC_{new} = (10.574)(58) = 613.31$/year

Payback calculations:
P = 6340 -2070 = $ 4270

Annual saving, F_t:
$F_t = AC_{conv.} - AC_{new} = 1{,}052.7$/year - 613.31/year = \$439.39/year

$$\sum_{i=1}^{n} F_t(1+i)^{-t} \geq P$$

Payback period, n, is found to be more than 100 years. Hence the proposed system employing solar panel regeneration is not an economically feasible option. Hence, there is no need for economical modeling.

0.6 Chapter 6

6.1

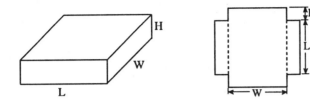

$$U = LWH \tag{6.1.1}$$
$$L = W \tag{6.1.2}$$
$$A = W(L + 2H) + 2LH = 40 \tag{6.1.3}$$

Substitute Eq. (6.1.2) into Eq. (6.1.3)

$$H = \frac{40 - W^2}{4W} \tag{6.1.4}$$

Substitute Eq. (6.1.2) and Eq. (6.1.4) into Eq. (6.1.1)

$$U = 10W - 1/4W^3 \tag{6.1.5}$$

Eq. (6.1.5) is the function to be maximized. The first derivative of Eq. (6.1.5) is
$\frac{dU}{dW} = 10 - 3/4W^2 = 0$, yields $W = \mp 3.65$, considering positive value of W, we have $\frac{d^2U}{dW^2} < 0$, the max is +3.65. Using Eq. 6.1.4 obtain H = 1.827, hence
U = LWH = (3.65)(3.65)(1.827) = 24.34 in^3.

6.2

$$LE = U(x,y,z) + \lambda_1 h_1(x,y,z) + \lambda_2 h_2(x,y,z) + \dots$$

Objective function is: $U = LWH$. Since we have an equality constraint of L=W, we can rewrite the objective function as: $U = L^2 H$.

Allowable area $= h_1 = WL + 2(WH) + 2(LH) = 80$ or
$h_1 = L^2 + 4LH - 80 = 0$, then
$LE = L^2 H + \lambda_1(L^2 + 4LH - 80)$

$$\frac{\partial LE}{\partial L} = 2LH + 2L\lambda_1 + 4H\lambda_1 = 0 \tag{6.2.1}$$

$$\frac{\partial LE}{\partial H} = L^2 + 4L\lambda_1 = 0 \tag{6.2.2}$$

$$\frac{\partial LE}{\partial \lambda_1} = L^2 + 4LH - 80 = 0 \tag{6.2.3}$$

From Eq. (6.2.2) $\lambda_1 = -\frac{L}{4}$ and substituting into Eq. (6.2.1) yields $H = L/2$. Using Eq. (6.2.3) yields
$L^2 + 4L(L/2) - 80 = 0$, therefore $L = \sqrt{80/3} = 5.16398$, then H=5.16398/2 = 2.58199.
Maximum volume of box with restriction of box area =80 in^2 is equal to:
$$V = (5.16398)^2(2.58199) = 68.85 \text{ in}^3$$

6.3
$U = x_1 + 4x_2$ subject to

$$x_1 \geq 0 \qquad (6.3.1)$$
$$x_2 \geq 0 \qquad (6.3.2)$$
$$3x_1 + 2x_2 \geq 6 \implies 3x_1 + 2x_2 = 6 \qquad (6.3.3)$$
$$x_1 + 2x_2 \geq 4 \implies x_1 + 2x_2 = 4 \qquad (6.3.4)$$

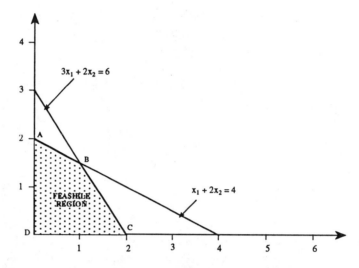

Best of maximum value will be at the intersections of the feasible region; select intersecting points for evaluation

Intersecting Points	$U(x) = x_1 + 4x_2$	
A(0,2)	8	BEST POINT
B(1,1.5)	7	
C(2,0)	2	
D(0,0)	0	

6.4

```
C....  THIS PROGRAM FINDS THE MINIMUM OF THE FUNCTION BY USING
C....  EQUAL INTERVAL SEARCH METHOD FOR EXAMPLE 6.4
C....
C      DELX : INITIAL STEP
C      EPSLON : ERROR CRITERION
C      X(1) : INITIAL POINT FOR STARTING CULCULATIONS
C      NCOUNT : NUMBER OF ITERATION
C....
       DIMENSION F(100),X(100)
       OPEN(6,FILE='ER3.OUT',STATUS='NEW')
       DELX = 0.5
       EPSLON = 0.001
       NCOUNT = 1
       X(1) = 10.
```

```fortran
C....
C.... CULCULATE THE VALUE OF THE FUNCTION FOR THE INITIAL VALUE OF X
C....
      CALL FUNC(1,X,F)
C....
      WRITE(6,*)' NCOUNT ','    X    ',' DELX ',' F '
      WRITE(*,*)' NCOUNT ','    X    ',' DELX ',' F '
      WRITE(6,*)' ——————— ',' ———————',' ——————- ',' ————'
      WRITE(*,*)' ——————— ',' ———————',' ——————- ',' ————'
      WRITE(*,30)NCOUNT,X(1),DELX,F(1)
      WRITE(6,30)NCOUNT,X(1),DELX,F(1)
      X(2) = X(1) + DELX
C....
C.... SEARCH THE EQUAL INTERVAL
C....
      DO 10 I = 2, 100000
      X(I) = X(I-1) + DELX
      CALL FUNC(I,X,F)
      WRITE(*,30)I,X(I),DELX,F(I)
      WRITE(6,30)I,X(I),DELX,F(I)
      IF(ABS(F(I)-F(I-1)) .LE. EPSLON)THEN
      X(I) = (X(I)+X(I-1)) / 2.
      CALL FUNC(I,X,F)
      WRITE(6,*)'VARIABLE OF X        = ',X(I)
      WRITE(6,*)'VALUE OF FUNCTION    = ',F(I)
      WRITE(6,*)'NUMBER OF ITERATION = ',I
      GO TO 20
C....    ELSE IF(F(I) .LE. F(I-1))THEN
C....    X(I+1) = X(I) + DELX
      ELSE IF(F(I) .GT. F(I-1))THEN
      DELX = -DELX / 2.
C....    X(I+1) = X(I) + DELX
      END IF
   10 CONTINUE
   20 WRITE(*,*)'VARIABLE OF X = ',X(I)
      WRITE(*,*)'VALUE OF FUNCTION = ',F(I)
      WRITE(*,*)'NUMBER OF ITERATION = ',I
   30 FORMAT(2X,I4,3X,3(F10.5,2X))
      STOP
      END
C....
C....    SUBROUTINE PROGRAM FOR FUNCTION
C....
      SUBROUTINE FUNC(I,X,F)
      DIMENSION F(1000),X(1000)
      F(I) = 3.0 * X(I)**2 + 1296.0 / X(I)
      RETURN
      END
```

6.5

```fortran
C..........................................................
C.... THIS PROGRAM FINDS THE MINIMUM FOR THE FUNCTION BY USING
C.... GOLDEN SECTION SEARCH METHOD FOR EXAMPLE 6.5
C..........................................................
C....
C.... XL : INITIAL LOWER VALUE OF INTERVAL
C.... XU : INITIAL UPPER VALUE OF INTERVAL
C.... UX1 : VALUE FOR LOWER INTERVAL
C.... UX2 : VALUE FOR UPPER INTERVAL
C.... EPSLON : ERROR CRITERION
C....
      OPEN(6,FILE='ER4.OUT',STATUS='NEW')
      XL = 1.
      XU = 11.
      EPSLON = 0.001
C....
C.... SEARCH THE MINIMUM VALUE OF THE FUNCTION
C....
      WRITE(*,*)' NCOUNT',' XL ',' XU ',' X1 ',
     .               ' X2 ',' UX1 ',' UX2 '
      WRITE(*,*)
      WRITE(6,*)' NCOUNT',' XL ',' XU ',' X1 ',
     .               ' X2 ',' UX1 ',' UX2 '
      WRITE(6,*)
C....
      DO 10 I = 1, 100000
      DIF = XU - XL
      X1 = XL + 0.618 * DIF
      X2 = XU - 0.618 * DIF
      CALL FUNC(X1,X2,UX1,UX2)
      WRITE(*,30) I,XL,XU,X1,X2,UX1,UX2
      WRITE(6,30) I,XL,XU,X1,X2,UX1,UX2
      IF(ABS(UX1-UX2) .LE. EPSLON) THEN
      X = (X1 + X2) / 2.
      CALL FUNC(X,X2,UX1,UX2)
      WRITE(6,*) 'VARIABLE OF X        = ',X
      WRITE(6,*) 'VALUE OF FUNCTION    = ',UX1
      WRITE(6,*) 'NUMBER OF ITERATION = ',I
      WRITE(*,*) 'VARIABLE OF X        = ',X
      WRITE(*,*) 'VALUE OF FUNCTION    = ',UX1
      WRITE(*,*) 'NUMBER OF ITERATION = ',I
      GO TO 20
      ELSE IF(UX1 .GT. UX2) THEN
      XU = X1
      ELSE
      XL = X2
      END IF
 10   CONTINUE
```

```
   30 FORMAT(2X,I4,6(F10.5,2X))
   20 STOP
      END
C....
C....    SUBROUTINE PROGRAM FOR FUNCTION
C....
      SUBROUTINE FUNC(X1,X2,UX1,UX2)
      UX1 = 3.0 * X1**2 + 1296.0 / X1
      UX2 = 3.0 * X2**2 + 1296.0 / X2
      RETURN
      END
```

0.7 Chapter 7

7.1
a) $t_{0.025,14} = 2.145$ b) $t_{0.05,29} = 1.699$
c) $P(-t_{\alpha/2} < t < t_{\alpha/2}) = 0.90$
$P(-t_{\alpha/2} < t < t_{\alpha/2}) = 1 - \alpha \Longrightarrow \alpha = 1 - 0.90 = 0.10$
Then, $t_{0.05,8} = 1.866$

7.2
a) $\chi^2_{0.025,12} = 23.337$ b) $\chi^2_{0.05,19} = 30.144$
c) $P(\chi^2 < \chi^2_\alpha) = 1 - \alpha = 0.995 \Longrightarrow \alpha = 0.005$
$\chi^2_{0.005,14} = 31.319$

7.3
a) $F_{0.1,7,14} = 2.19$ b) $F_{0.05,3,28} = 2.95$ c) $F_{0.025,40,60} = 1.74$

7.4
$\bar{x} = \frac{\sum_{i=1}^{n} x_i}{n} = \frac{2.50+2.52+\cdots+2.55}{10} = 2.5075$
$S^2 = \frac{\sum_{i=1}^{n}(x_i-\bar{x})^2}{n-1} = \left[\frac{(2.50-2.5075)^2+(2.52-2.5075)^2+\cdots+(2.55-2.5075)^2}{8-1}\right] = 0.0006786$
$\nu = 8 - 1 = 7$, since confidence level is 90%, we choose $\alpha=0.10$, then $\chi^2_{0.05,7} = 14.067$ and $\chi^2_{0.95,7} = 2.167$.
$\frac{(n-1)S^2}{\chi^2_{\alpha/2,\nu}} \leq \hat{\sigma}^2 \leq \frac{(n-1)S^2}{\chi^2_{1-\alpha/2,\nu}}$
$\frac{(7)(0.0006786)}{14.067} \leq \hat{\sigma}^2 \leq \frac{(7)(0.0006786)}{2.167} \Longrightarrow 0.00034 \leq \hat{\sigma}^2 \leq 0.0022$

7.5
Use Chi-square distribution for confidence level

$$\frac{(n-1)S^2}{\chi^2_{\alpha/2,\nu}} \leq \hat{\sigma}^2 \leq \frac{(n-1)S^2}{\chi^2_{1-\alpha/2,\nu}} \tag{7.5.1}$$

$n = 28$, $\nu = n - 1 = 28 - 1 = 27$, $\alpha = 0.05$, $\alpha/2 = 0.025$, $1 - \alpha/2 = 0.975$,
$\chi^2_{\alpha/2,\nu} = \chi^2_{0.025,27} = 43.19$, $\chi^2_{0.975,27} = 14.57$
Substituting into Eq. (7.5.1)

$$\frac{(28-1)2.9}{43.19} \leq \hat{\sigma}^2 \leq \frac{(28-1)2.9}{14.57}, \quad \text{or}$$

$1.81 \leq \hat{\sigma}^2 \leq 5.37$ and
$1.345 \leq \hat{\sigma} \leq 2.317$

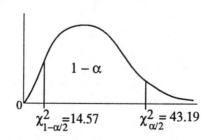

Since desired $\hat{\sigma}=3$ kips/in with 95 percent level of confidence does not fall within the calculated limits, the data doesn't support the claim.

7.6

Since the sample size is less than 30, we use t distribution. Using data given in Table A.6 find $\bar{x}_1 = 112.5$ and $\bar{x}_2 = 124.5$. Hence,

$S_2^2 = \frac{(120-112.5)^2+(118-112.5)^2+\cdots+(121-112.5)^2}{8-1} = 54.85$

$S_1^2 = \frac{(121-124.5)^2+(125-124.5)^2+\cdots+(135-124.5)^2}{8-1} = 41.99$

The Pooled estimator is

$S_p^2 = \frac{(n_1-1)S_1^2+(n_2-1)S_2^2}{n_1+n_2-2}$

$S_p^2 = \frac{(8-1)54.85+(8-1)41.99}{8+8-2} = 48.42$, then $S_p = 6.96$

$100(1-\alpha)$ percent two-sided confidence interval for the difference in means $(\mu_1 - \mu_2)$ is

$(\bar{x}_1 - \bar{x}_2) - t_{\alpha/2,n_1+n_2-2}S_p\sqrt{1/n_1 + 1/n_2} \leq (\mu_1 - \mu_2) \leq (\bar{x}_1 - \bar{x}_2) + t_{\alpha/2,n_1+n_2-2}S_p\sqrt{1/n_1 + 1/n_2}$

Substituting

$(112.5-124.5) - (2.145)(6.96)\sqrt{1/8+1/8} \leq (\mu_1-\mu_2) \leq (112.5-124.5)+(2.145)(6.96)\sqrt{1/8+1/8}$,

or

$-19.465 \leq (\mu_1 - \mu_2) \leq -4.5354$

Since the confidence interval doesn't include $\mu_1 - \mu_2 = 0$, there is statistical difference in the performance of two manufacturing processes at the 95 percent level of confidence.

7.7

Let $n=6$ be the number of injectors and $p = 1/200 = 0.005$ be the probability of the injector failing, then

$q = 1 - p = 1 - 0.005 = 0.995$

The probability of the engine not failing is when not more than $x=3$ injectors fail is

$$p(x) = \sum_{x=0}^{r} \binom{n}{x} p^x q^{n-x} \quad for \quad x = 0, 1, 2 \cdots n \tag{7.7.1}$$

where

$\binom{n}{x} = \frac{n!}{x!(n-x)!}$, substituting

$p(3) = \frac{6!}{3!3!}(0.005)^3(0.995)^3 + \frac{6!}{2!4!}(0.005)^2(0.995)^4 + \frac{6!}{1!5!}(0.005)^1(0.995)^5 + \frac{6!}{0!6!}(0.005)^0(0.995)^6$

$p(3) = 0.99999999$

Therefore the probability of the engine failing is

$1 - p(3) = 1 - 0.99999999 = 9.261 \times 10^{-9}$

7.8

$n=10$, $\bar{x}=4.9 \times 10^6$ cycles, $S=4.1 \times 10^5$ cycles.

Step 1. Set the null hypothesis

$H_0 \Longrightarrow \mu = \mu_H = 6 \times 10^6$ cycles

$H_1 \Longrightarrow \mu \neq \mu_H \neq 6 \times 10^6$ cycles

Step 2. Assume a level of significance, $\alpha = 0.01$ (two sided)

Step 3. Size of sample, $n=10 < 30$

Step 4. Since $n = 10 < 30$ and assuming that x is normally distributed, the following t distribution is used

$t = \frac{\bar{x}-\mu}{S/\sqrt{n}}$

$t = \frac{4.9\times 10^6 - 6\times 10^6}{4.1\times 10^5 / \sqrt{10}} = -8.48$

Step 5. Assuming the two sided region with $\alpha=0.005$ and the degree of freedom, $\nu=10-1=9$, the value of t for the critical region are found to be $t = \mp 3.250$ (from Table C-2, Appendix C)

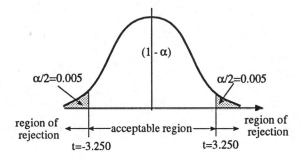

Since $t_{\bar{x}}=-8.48$ is not in the acceptable region, reject the hypothesis, H_0, and decide that no significant evidence is available to support that the new material increases the fatigue life at $\alpha=0.01$.

7.9

$\mu=68$ psi $\implies$ pressure for normal operation
$\mu=64$ psi $\implies$ pressure for abnormal operation
$\hat{\sigma}=2.5$ psi

Decision	STATE	
	Abnormal Operation p(abnormal)=0.70	Normal Operation p(normal)=0.30
Continue operation	r=1	R=0
Don't continue operation	r=0	R=5

Expected loss when operation is continued = $(1)(0.70) + (0)(0.30) = 0.70$
Expected loss when operation is not continued = $(0)(0.70) + (5)(0.30) = 1.5$

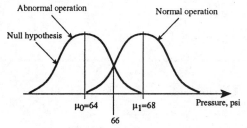

Fig. 7.9.1 Decision making curve

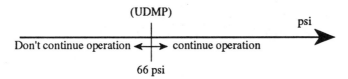

Fig. 7.9.2. Critical region for decision making

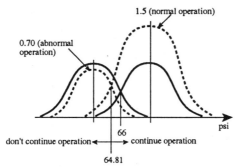

Fig. 7.9.3. Modified decision making curves

Results show that the expected loss is less when operation is continued. Hence, continue to operate the pump is correct decision.

UDMP $= \frac{\mu_0 + \mu_1}{2} = \frac{64+68}{2} = 66$ psi

The change in UDMP when probabilities are taken into consideration is

$\Delta'_{UDMP} = \frac{\hat{\sigma}^2}{\mu_1 - \mu_0} \times ln \frac{P(\text{Null hypothesis is true})}{P(\text{Null hypothesis is false})}$

$\Delta'_{UDMP} = \frac{2.5^2}{68-64} \times ln \frac{0.70}{0.30} = 1.3239$

The change in UDMP when the regret is taken into consideration is

$\Delta''_{UDMP} = \frac{\hat{\sigma}^2}{\mu_1 - \mu_0} \times ln \frac{P(\text{Regrets of TYPE I Error})}{P(\text{Regrets of TYPE II Error})}$

$\Delta''_{UDMP} = \frac{2.5^2}{68-64} \times ln \frac{1}{5} = -2.5147$ psi

The net change in the decision making point is 1.3239 - 2.5147 = - 1.1908. Hence, the resulting decision making point is 66 -1.1908 $\approx$ 64.81. Since the modified decision making point is 64.81 psi and the observed oil pressure is 65 psi, the best decision would be to continue operation.

The probability of a TYPE I error, α, from the following figure is

$\alpha = 1 - p(z < \frac{x-\mu_0}{\hat{\sigma}/\sqrt{n}}) = 1 - p(z < \frac{64.81-64}{2.5/\sqrt{5}})$

$\alpha = 1 - p(z < 0.72) \approx 0.2358$

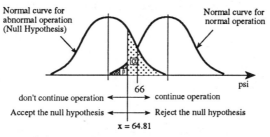

Fig. 7.9.4. Probability of TYPE I and TYPE II errors

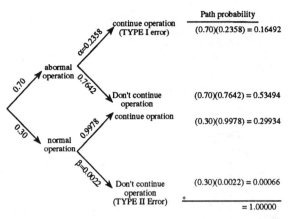

Fig. 7.9.5. Tree diagram

The probability of a TYPE II error, β, is

$\beta = p(z < \frac{x-\mu_1}{\hat{\sigma}/\sqrt{n}}) = p(z < \frac{64.81-68}{2.5/\sqrt{5}})$

$\beta = p(z < -2.85) = 0.0022$

These areas are shown in the above figure.
Since the observed oil pressure, 65 psi, within the critical region to the right of the decision making point of 64.81 psi, operation should be continued. Thus, reject the Null Hypothesis.

From the tree diagram the path of probability for a TYPE I error is 0.16492 and for TYPE II error is 0.00066. Thus, expected loss is

$(1)(0.16492) + (5)(0.00066) = 0.16822$

Previously calculated smallest loss was 0.70. After the correction, expected loss is reduced by the factor of 4.

7.10 Up-and-down test for sensitivity of explosive

	Test No:	1	2	3	4	5	6	7	8	9	10	11	12	13	14	15	16	17	18	X	0
																				X	0
	3.0								x		x		x							3	0
Normalized height	2.6						0		0		0	x								1	3
	2.2			x		0							x		x		x			4	1
	1.8		0		0									0			0		0	0	5
	1.4	0																		0	1
x indicates explosion																					
0 indicates nonexplosion																				8	10

Since there are fewer successes than the failures, x is used in the analysis. Also note that since the explosion did not occur at the 1.4 and 1.8 levels, data at these levels are eliminated.

Normalized hight	i	n_i	$i\, n_i$	$i^2\, n_i$
3.0	2	3	6	12
2.6	1	1	1	1
2.2	0	4	0	0
		N=8	A=7	B=13

Sample mean is

$$\bar{x} = D_0 + I\left(\frac{A}{N} - \frac{1}{2}\right) = 2.2 + 0.4\left[\frac{7}{8} - \frac{1}{2}\right] = 2.35$$

Standard deviation of the sample is

$$S = 1.620 \times I\left(\frac{NB-A^2}{N^2} + 0.029\right) = 1.620 \times 0.4\left(\frac{8 \times 13 - 7^2}{8^2} + 0.029\right) = 0.576$$

7.11

Up-and-down test for sensitivity of explosive

Test No:	1	2	3	4	5	6	7	8	9	10	11	12	13	14	15	16	17	18	x	0
3.0	x																		1	0
2.6		x		x										x					3	0
2.2			0		x		x		x		x		0		x				5	2
1.8						0		0		0		0				x		0	1	5
1.4																	0		0	1
																			10	8

x indicates explosion
0 indicates nonexplosion

Test results show that there are 10 explosions and 8 no explosions (failure). Hence the failures are used to estimate the parameters. Note that failures appear on three levels (1.4, 1.8, 2.2). The levels 2.6 and 3.0 are eliminated.

Normalized hight	i	n_i	$i\,n_i$	$i^2 n_i$
2.2	2	2	4	8
1.8	1	5	5	5
1.4	0	1	0	0
		N=8	A=9	B=13

Sample mean is

$$\bar{x} = D_0 + I\left(\frac{A}{N} + \frac{1}{2}\right) = 1.4 + 0.4\left[\frac{9}{8} + \frac{1}{2}\right] = 2.05$$

Standard deviation of the sample is

$$S = 1.620 \times I\left(\frac{NB-A^2}{N^2} + 0.029\right) = 1.620 \times 0.4\left(\frac{8 \times 13 - 9^2}{8^2} + 0.029\right) = 0.25$$

7.12

Two-factor L_4 orthogonal array

Col. → No.↓	Roughness 1	Speed 2	Error 3	Observations, Y_i
1	1	1	1	0.050
2	1	2	2	0.054
3	2	1	2	0.061
4	2	2	1	0.069

High value of speed=5.0 rpm
Low value of speed=1.5 rpm
High value of roughness=150 μin
Low value of roughness=75 μin

The solution of the problem is summarized in ANOVA table. The percent contribution shows that the factor roughness contributes the most to the variation observed in the experiment and this agrees with the F test ratio.

ANOVA summary for two factorial experiment

Source	Deg-of-freedom (ν)	Sum of squares (SS_i)	Mean squares (MS_i)	F Test	Percent of variance
rough	1	0.000169	0.000169	42.25	78.95
speed	1	0.000036	0.000036	9	15.31
Error	1	0.000004	0.000004		5.74

7.13

$$F_{speed} = \frac{MS_{speed}}{MS_E} = \frac{45.125}{0.125} = 361$$

F values from the table:

$F_{\alpha=0.25, \nu_1=1, \nu_2=1} = 5.83$

$F_{\alpha=0.10, \nu_1=1, \nu_2=1} = 38.86$

$F_{\alpha=0.05, \nu_1=1, \nu_2=1} = 161.4$

$F_{\alpha=0.025, \nu_1=1, \nu_2=1} = 647.8$

Above results show that the minimum confidence level is 95%, and $\alpha=0.05$.

7.14

ANOVA summary for spin rinse drying experiment

Source	Deg-of-freedom (ν)	Sum of squares (SS_i)	Mean squares (MS_i)	F Test	Percent of variance
process	1	128.0	128.0	0.0385	0.00
inspection	1	11250	11250	3.3880	7.91
spin	1	65884.5	65884.5	19.8462	62.44
proc. × inspec.	1	7200	7200	2.1683	3.87
proc. × spin	1	84.5	84.5	0.0254	0.0
inspec. × spin	1	12324.5	12324.5	3.7116	8.99
Error	1	3320.5	3320.5		16.79

The solution of the problem is summarized in ANOVA table. The percent contribution shows that the factor spin contributes the most to the variation observed in the experiment and this agrees with the F test ratio.

7.15

ANOVA summary for spin rinse drying lumped experiment

Source	Deg-of-freedom (ν)	Sum of squares (SS_i)	Mean squares (MS_i)	F Test	Percent of variance
process	1	128.0	128.0	0.0223	0.00
inspection	1	11250	11250	1.9625	5.51
spin	1	65884.5	65884.5	11.4934	60.04
Error	4	22929.5	5732.4		34.45

From the result of F test and percent contribution, it is clear that the spin is by far the largest contributing factor to particulates. Note that the increase of contribution in error compare to problem 7.14.

7.16

ANOVA summary for arc chamber lumped experiment

Source	Deg-of-freedom (ν)	Sum of squares (SS_i)	Mean squares (MS_i)	F Test	Percent of variance
arc current	1	0.2700	0.2700	46.66	58.33
extraction current	1	0.1378	0.1378	23.82	29.14
pressure	1	0.02205	0.02205	3.81	3.59
Error	4	0.02315	0.005786		8.94

ANOVA table shows that the contribution of pressure is insignificant. The effect of arc current is maximum to the experimental design. Of course, the error will be reduced if the interactions are considered.

7.17

$VIS = \alpha * exp(\beta * T)$

$Log(VIS) = Log(\alpha) + \beta * T$

The value of $log(\alpha)$ represents the intercept, and β represents the gradient. The $log(VIS)$ is modeled with temperature in simple regression **GLM** procedure as

```
TITLE 'ABSOLUTE VISCOSITY OF CELD SOLUTION (45%)';
OPTIONS LS=72 PS=66 NODATE;
FILENAME IN 'WORK$AREA:[kvkvs]VIS.DAT';
DATA ONE
INFILE IN;
INPUT T 2-7 VP 14-19;
DATA TWO;
SET ONE;
LVP = LOG(VIS);
PROC GLM;
MODEL LVP = T;
PROC PRINT;
```

In interpreting the results, the antilog of the intercept must be taken to give the actual value of α. From the output table substituting the intercept and gradient into equation, gives

$Log(VIS) = 3.27478844 - 0.03513031 \times T$

SAS output

Temperature	Actual	Predict
15.55	15.713	15.3101
21.11	12.786	12.5936
26.66	10.548	10.3627
32.22	8.424	8.5240
37.00	6.871	7.0141
43.33	5.439	5.7695
48.88	4.515	4.7475
54.44	3.900	3.9051
60.00	3.494	3.2122
INTERCEPT	...	3.27478844
GRADIENT	...	-0.03513031

Taking the antilog,

$(VIS) = e^{3.27478844} e^{-0.03513031 \times T}$

$(VIS) = \alpha e^{-0.03513031 \times T}$

calculating $\alpha = e^{3.27478844} = 26.437$ and rewriting the predicted equation as

$(VP) = 26.437 e^{-0.03513031 \times T}$

0.8 Chapter 8

8.1
$P(A) = 0.6$, $P(B) = 0.2$, and $P(A\&B) = 0.15$ then

$P(A) \times P(B) = 0.2 \times 0.6 = 0.12 \neq P(A\&B)$

Therefore A and B are not independent.

8.2
Probability of blue pen drawn depends on the probability of red pen drawn. Assume event R is red pen and event B is blue pen. Total number of pens are 20.

P(first pen red *and* second pen blue) = P(first pen red) • P(second pen is blue; given that first pen is red), or

$P(R \text{ and } B) = P(R) \bullet P(B|R)$

$P(R \text{ and } B) = \frac{4}{20} \times \frac{9}{19} = \frac{9}{95}$

8.3
If A represents yellow for the first (outer) ball and B represents yellow for the second (inner) ball, probability of either the first or second ball showing yellow can be calculated by

$P(A + B) = P(A) + P(B) - P(AB)$

where

$P(A) = 2/5, P(B) = 2/5$

Since both events are independent

$P(AB) = P(A) \times P(B)$, then

$P(A + B) = \frac{2}{5} + \frac{2}{5} - \frac{2}{5} \times \frac{2}{5} = \frac{16}{25}$

8.4
Assume A_1 is the lot with 50% reliability and A_2 is the lot with 80% reliability. Let B denote the event failure (doesn't explode). Since we have two lots, the probability of selecting lot A_1 and A_2 is 50% (each lot has equal chance of being chosen). Then

$P(A_1) = 0.50$, and $P(A_2) = 0.50$

Since reliability of lot A_1 is 50%, probability of failure, $P(B|A_1) = 1 - R_1 = 1 - 0.50 = 0.50$

Similarly, $P(B|A_2) = 1 - R_2 = 1 - 0.80 = 0.20$

Using Bayes' theorem, probability that the selected explosive material was taken from lot A_1 is

$P(A_1|B) = \frac{P(B|A_1)P(A_1)}{P(A_1)P(B|A_1)+P(A_2)P(B|A_2)} = \frac{0.5 \times 0.5}{0.5 \times 0.5 + 0.5 \times 0.2} = 0.714$

and the probability that the selected explosive material was taken from lot A_2 is

$P(A_2|B) = \frac{P(B|A_2)P(A_2)}{P(A_1)P(B|A_1)+P(A_2)P(B|A_2)} = \frac{0.2 \times 0.5}{0.5 \times 0.5 + 0.5 \times 0.2} = 0.286$

8.5
Assume that the bearing operation starts at $t=0$, then

1) $R(t) = exp[-(\frac{t}{\theta})^b] = exp[-(\frac{400}{1200})^{1.25}] = exp(-0.2533) = 0.776$

2) $MTTF = \theta\Gamma(\frac{1}{b} + 1) = 1200\Gamma(\frac{1}{1.25} + 1) = (1200)\Gamma(1.8) = (1200)(1.37) = 1644$ hr.

3) $h(t) = \frac{bt^{b-1}}{\theta^b} = \frac{(1.25)(400)^{1.25-1}}{(1200)^{1.25}} = 0.00079$ failure/hr

---

8.6
$R^2 = 0.985$

$R = 0.992$

8.7
From the statement of the problem it is clear that the system is connected in parallel, hence

$R_p = 1 - \prod_{k=1}^{n}(1 - R_k) = 1 - \left\{(1 - R_1)(1 - R_2)\right\} = 1 - \left\{(1 - 0.85)(1 - 0.89)\right\} = 0.9835$

8.8
From the statement of the problem it is clear that the system is connected in series, hence

$R_s = \prod_{k=1}^{n} R_k = R_1 R_2 R_3 .... R_n$

$R_s = (0.96)^2 = 0.9216$

8.9
Since the system will fail if both compressor fail, system is parrallel, then

$R_p = 1 - (1 - R_k)^k = 1 - (1 - R_k)^2$

For constant λ, $R = e^{-\lambda t}$, then

$R_p = 1 - (1 - e^{-\lambda t})^2 = 1 - \left[1 - e^{-(0.00015)(1500)}\right]^2$

$R_p = 1 - \left(1 - e^{-0.225}\right)^2 = 0.96$

$MTTF = \frac{1}{\lambda} = \frac{1}{0.00015} = 6666.7$ hr.

8.10
$\phi_1 = \frac{1-R_1}{R_1} = \frac{1-0.85}{0.85} = 0.1764706$

$\phi_2 = \frac{1-R_2}{R_2} = \frac{1-0.89}{0.89} = 0.1235955$

System parameter, ϕ_s is

$\phi_s = \phi_1 \phi_2 = (0.1764706)(0.1235955) = 0.021811$

Then the reliability of the system is

$R_s = \frac{1}{1+\phi_s} = \frac{1}{1+0.021811} = 0.97865$

8.11
Using classical method, the reliability of the system is

$R_{sp} = \prod_{k=1}^{n}[1 - (1 - R_k)^{Y_k}]$

$R_{sp} = [1 - (1 - 0.90)^2][1 - (1 - 0.85)^3][1 - (1 - 0.80)^4][1 - (1 - 0.70)^5] = 0.982687$

Using parametric approach the reliability of the system is

$\phi_{sp} = \sum_{k=1}^{n} \phi_k^{Y_k} = \phi_1^2 + \phi_2^3 + \phi_3^4 + \phi_4^5$

$\phi_1 = \frac{1-R_1}{R_1} = \frac{1-0.90}{0.90} = 0.1111$

$\phi_2 = \frac{1-R_2}{R_2} = \frac{1-0.85}{0.85} = 0.1764706$

$\phi_3 = \frac{1-R_3}{R_3} = \frac{1-0.80}{0.80} = 0.25$

$\phi_4 = \frac{1-R_4}{R_4} = \frac{1-0.70}{0.70} = 0.4285714$

$\phi_{sp} = (0.1111)^2 + (0.1764706)^3 + (0.25)^4 + (0.4285714)^5 = 0.0362034$

$R_{sp} = \frac{1}{1+\phi_{sp}} = \frac{1}{1+0.0362034} = 0.965$

8.12

Table 8.12.1   3 stage redundant system

Stage	Cost per unit	Reliability
1	$C_1=2$	$R_1=0.63$
2	$C_2=5$	$R_2=0.72$
3	$C_3=7$	$R_3=0.86$

For stage 1.

$\phi_1 = \frac{1-R_1}{R_1} = \frac{1-0.63}{0.63} = 0.5873$

$K_i = \frac{ln\phi_1 C_i}{ln\phi_i C_1}$ for $i=1$, $K_1 = \frac{ln\phi_1 C_1}{ln\phi_1 C_1} = \frac{ln 0.5873 \times 2}{ln 0.5873 \times 2} = 1$

For stage 2.

$\phi_2 = \frac{1-R_2}{R_2} = \frac{1-0.72}{0.72} = 0.3888$

$K_2 = \frac{ln\phi_1 C_2}{ln\phi_2 C_1} = \frac{ln 0.5873 \times 5}{ln 0.3888 \times 2} = 1.4084499$

Similarly, for stage 3 we have $\phi_3=0.1628$ and $K_3=1.026187$

$K = \sum K_i = 1 + 1.408 + 1.026 = 3.434$ $\phi_L = \frac{1-R_L}{R_L} = \frac{1-0.95}{0.95} = 5.2631 \times 10^{-2}$

$\lambda = -\frac{C_1}{S \times ln\phi_1} = -\frac{2}{1.53266 \times 10^{-2} \times ln 0.5873} = 245.1847$

Find the smallest number that will satisfy the following inequality equation

$\lambda \frac{\phi_i^{Y_i}}{[1+\phi_i]^{(1+Y_i)}} < C_i$ assume $\Delta\lambda=40$ with $Y_1=0$ and $\lambda = 245.1847$ start iteration

Consider left hand side (LHS) of the inequality equation

$\lambda \frac{\phi_1^{Y_1}}{[1+\phi_1]^{(1+Y_1)}} = \frac{245.1847 \times 0.5873^0}{[1+0.5873]^{(1+0)}} = 154.4665$

Since $154.4665 > C_1 = 2$, inequality is not satisfied, hence, increment Y_1 by 1 and recalculate the LHS.

$\lambda \frac{\phi_1^1}{[1+\phi_1]^{(1+1)}} = \frac{245.1847 \times 0.5873^1}{[1+0.5873]^2} = 57.1525$

When $Y_i = 5$ we have

$\lambda \frac{\phi_1^5}{[1+\phi_1]^{(1+5)}} = \frac{245.1847 \times 0.5873^5}{[1+0.5873]^6} = 1.07112$

$1.07112 < C_1 = 2$ which satisfies the inequality equation. Thus the number of components in the first stage is 5. Similarly, this same calculations can be performed for stage 2 and stage 3. Using same $\lambda=245.1847$, the number of components for stages 2 and 3 can be found to be 3 and 2, respectively. The system total cost, C_s is

$C_s = (5 \times 2) + (3 \times 5) + (2 \times 7) = 39.0$

Calculate the system reliability

$R_{sp} = \prod_{k=1}^{n}[1 - (-R_k)^{Y_k}]$
$= [1 - (1 - 0.63)^5][1 - (1 - 0.72)^3][1 - (1 - 0.86)^2] = 0.9527$

Since $R_L \leq R_{sp}$, we stop iteration. If $R_L \geq R_{sp}$, then, we continue iteration with $\lambda = \lambda_1 - \Delta\lambda = 245.1847 - 40 = 205.1847$ until we have the constraints of $R_L \leq R_{sp}$ is satisfied.

The output of the computer program in Appendix D-3 is shown below:

NUMBER OF ELEMENTS	STAGES
5.00	1
3.00	2
2.00	3

COST = 39.0000000000000

RELIABILITY = 0.9522290179226

8.13

Table 8.13-1

Stage	Cost per unit	Volume	Reliability
1	$C_1=2$	$V_1=3$	$R_1=0.63$
2	$C_2=5$	$V_2=2$	$R_2=0.72$
3	$C_3=7$	$V_3=4$	$R_3=0.86$

Cost constraint = 40 and volume constraint = 40.

$$\lambda = \lambda_1 + \lambda_2 \tag{8.13.1}$$

is calculated from the following nonlinear simultaneous equations:

$$C_L = -\sum_{k=1}^{n} a_k[ln(a_k\lambda_1 + b_k\lambda_2)] \tag{8.13.2}$$

$$V_L = -\sum_{k=1}^{n} b_k[ln(a_k\lambda_1 + b_k\lambda_2)] \tag{8.13.3}$$

where

$$a_i = -\frac{C_i}{ln\phi_i} \text{ and } b_i = -\frac{V_i}{ln\phi_i} \tag{8.13.4}$$

Stage 1.

$\phi_1=0.5873$ then $a_1 = -\frac{2}{ln 0.5873} = 3.7578$ and $b_1 = -\frac{3}{ln 0.5873} = 5.6368$. Similarly, a_i and b_i are found for the other stages and the results are shown in Table 8.13-2

Newton-Ralphson method can be used to calculate λ_1 and λ_2. Iterations are started by assuming initial λ_1 and λ_2 as 0.001 and 0.001, respectively. The acceptable error is assumed to be 1%.

Newton-Ralphson method suggests that

Table 8.13-2

Stage	Cost per unit	Volume	ϕ_i	a_i	b_i
1	$C_1=2$	$V_1=3$	0.5873	3.7578	5.6368
2	$C_2=5$	$V_2=2$	0.3888	5.2940	2.1176
3	$C_3=7$	$V_3=4$	0.1628	3.8561	2.2035

$$x_{i+1} = x_i - \frac{u_i \frac{\partial v_i}{\partial y} - v_i \frac{\partial u_i}{\partial y}}{\frac{\partial u_i}{\partial x}\frac{\partial v_i}{\partial y} - \frac{\partial u_i}{\partial y}\frac{\partial v_i}{\partial x}} \qquad (8.13.5)$$

$$y_{i+1} = y_i + \frac{u_i \frac{\partial v_i}{\partial x} - v_i \frac{\partial u_i}{\partial x}}{\frac{\partial u_i}{\partial x}\frac{\partial v_i}{\partial y} - \frac{\partial u_i}{\partial y}\frac{\partial v_i}{\partial x}} \qquad (8.13.6)$$

In the above equations, we let $x = \lambda_1$, $y = \lambda_2$, and

$$u = C_L + \sum_{k=1}^{n} a_k[ln(a_k\lambda_1 + b_k\lambda_2)] \qquad (8.13.7)$$

$$v = V_L + \sum_{k=1}^{n} b_k[ln(a_k\lambda_1 + b_k\lambda_2)] \qquad (8.13.8)$$

$$\frac{\partial u}{\partial x} = \frac{\partial u}{\partial \lambda_1} = \sum_{k=1}^{n} a_k^2 \frac{1}{a_k\lambda_1 + b_k\lambda_2} \qquad (8.13.9)$$

$$\frac{\partial u}{\partial y} = \frac{\partial u}{\partial \lambda_2} = \sum_{k=1}^{n} a_k b_k \frac{1}{a_k\lambda_1 + b_k\lambda_2} \qquad (8.13.10)$$

$$\frac{\partial v}{\partial x} = \frac{\partial v}{\partial \lambda_1} = \sum_{k=1}^{n} a_k b_k \frac{1}{a_k\lambda_1 + b_k\lambda_2} \qquad (8.13.11)$$

$$\frac{\partial v}{\partial y} = \frac{\partial v}{\partial \lambda_2} = \sum_{k=1}^{n} b_k^2 \frac{1}{a_k\lambda_1 + b_k\lambda_2} \qquad (8.13.12)$$

Iteration starts with guessed λ_1 and λ_2 to calculate Equations (8.13.12), (8.13.11), (8.13.10), (8.13.9), (8.13.8), and (8.13.7) and substitute into Equations (8.13.5) and (8.13.6) to calculate $\lambda_2(i+1)$ and $\lambda_1(i+1)$. This iteration is continued until the error criterion is satisfied.

$\lambda_1=0.001$ and $\lambda_2=0.001$

$\frac{\partial u_1}{\partial x} = \frac{\partial u}{\partial \lambda_1} = \sum_{k=1}^{n} a_k^2 \frac{1}{a_k\lambda_1 + b_k\lambda_2} = (3.7578)^2 \frac{1}{[(3.7578)(0.001)+(5.6368)(0.001)]}$

$+(5.2940)^2 \frac{1}{[(5.2940)(0.001)+(2.1176)(0.001)]} + (3.8561)^2 \frac{1}{[(3.8561)(0.001)+(2.2035)(0.001)]} = 7738.41$

Similarly, $\frac{\partial u_1}{\partial y} = 5169.53$, $\frac{\partial v_1}{\partial x} = 5169.53$, $\frac{\partial v_1}{\partial y} = 4788.38$

Then

$u = C_L + \sum_{k=1}^{n} a_k[ln(a_k\lambda_1 + b_k\lambda_2)] = 40 + 3.7578[ln(3.7578 \times 0.001 + 5.6368 \times 0.001)]$

$+5.2940[ln(5.2940 \times 0.001 + 2.1176 \times 0.001)] + 3.8561[ln(3.8561 \times 0.001 + 2.2035 \times 0.001)] = -23.195$

Similarly, $v_1 = -7.947$ can be obtained.

Using Equation (8.13.5)

$$x_{i+1} = \lambda_{i+1} = x_1 - \frac{u_1 \frac{\partial v_1}{\partial y} - v_1 \frac{\partial u_1}{\partial y}}{\frac{\partial u_1}{\partial x}\frac{\partial v_1}{\partial y} - \frac{\partial u_1}{\partial y}\frac{\partial v_1}{\partial x}}$$

$$\lambda_{i+1} = 0.001 - \frac{(-23.195)(4788.38) - (-7.947)(5169.53)}{(7738.4)(4788.38) - (5169.53)(5169.53)} = 7.77 \times 10^{-3}$$

Similarly, the value of $\lambda_2 = -4.65 \times 10^{-3}$ can be obtained. Repeat the above calculations with the new values of $\lambda_1 = 7.77 \times 10^{-3}$ and $\lambda_2 = -4.65 \times 10^{-3}$

After 5 iterations λ_1 converges to 3.641×10^{-2} and λ_2 converges to -2.349×10^{-2}. Then

$$\lambda = \lambda_1 + \lambda_2 = 3.641 \times 10^{-2} + (-2.349 \times 10^{-2}) = 1.292 \times 10^{-2}$$

Assuming $\Delta \lambda = 0.002$ and $y_1 = 0.0$ satisfy the following inequality equation

$$\frac{\phi_i^{Y_i}}{[1+\phi_i]^{(1+Y_i)}} < \lambda(g_1 C_i + g_2 V_i)$$

where

$g_1 = \frac{\lambda_1}{\lambda} = \frac{3.641 \times 10^{-2}}{1.292 \times 10^{-2}} = 2.8181115$ and $g_2 = \frac{\lambda_2}{\lambda} = \frac{-2.349 \times 10^{-2}}{1.292 \times 10^{-2}} = -1.8181115$

For the first stage $C_1 = 2$ and $V_1 = 3$ we have

$\lambda(g_1 C_1 + g_2 V_1) = 1.292 \times 10^{-2}[(2.8181115)(2) + (-1.8181115)(3)] = 2.35 \times 10^{-3}$, and

$$\frac{\phi_i^{Y_i}}{[1+\phi_i]^{(1+Y_i)}} = \frac{(0.5873)^0}{(1+0.5873)^{(0+1)}} = 0.6300$$

Since $0.6300 > 2.35 \times 10^{-3}$ doesn't satisfy the inequality equation. Therefore, increment y_1 by 1 and repeat the same calculations until the inequality is satisfied. When $y_1 = 6$

$$\frac{\phi_i^{Y_i}}{[1+\phi_i]^{(1+Y_i)}} = \frac{(0.5873)^6}{(1+0.5873)^7} = 0.0016164$$

$1.6164 \times 10^{-3} < 2.35 \times 10^{-3}$ satisfies the inequality equation. When the same calculations are repeated for the second and third stages, we have the following results

NUMBER OF ELEMENTS	STAGES
6	1
2	2
1	3

$R_{sp} = \prod_{k=1}^{n}[1 - (1 - R_k)^{Y_k}]$
$= [1 - (1 - 0.63)^6][1 - (1 - 0.72)^2][1 - (1 - 0.86)^1] = 0.7905$

$C' = 2 \times 6 + 5 \times 2 + 7 \times 1 = 29$

$V' = 3 \times 6 + 2 \times 2 + 4 \times 1 = 26$

Since $C_L > C'$ and $V_L > V'$, update $\lambda = \lambda - \Delta\lambda = 1.293 \times 10^{-2} - 0.002 = 0.01092$

next iteration with $\lambda = 0.0109 - 0.002 = 0.00892$

and next iteration with $\lambda = 0.00892 - 0.002 = 0.00692$. For $\lambda = 0.00692$ we have the following result

NUMBER OF ELEMENTS	STAGES
7	1
2	2
2	3

COST = 38.0000000000000

VOLUME = 33.0000000000000

RELIABILITY = 0.90267

Next calculation will be with $\lambda = 0.00692 - 0.002 = 0.00492$. The calculated results are

NUMBER OF ELEMENTS	STAGES
7	1
3	2
2	3

COST = 43.0000000000000

VOLUME = 35.0000000000000

RELIABILITY = 0.95796797

Since $C_L = 40 < C' = 43$, this result is not acceptable. Therefore, the optimum solution is the previous one.

8.14
The following results are obtained by using the computer programming given in Appendix D-3

NUMBER OF ELEMENTS	STAGES
2	1
3	2
4	3
5	4

COST = 25.0000000000000

RELIABILITY = 0.9929752894959

8.15
The following results are obtained by using the computer programming given in Appendix D-2

NUMBER OF ELEMENTS	STAGES
2	1
3	2
4	3
5	4

COST = 33.5000000000000

RELIABILITY = 0.9833085262948

8.16
The following results are obtained by using the computer programming given in Appendix D-4

NUMBER OF ELEMENTS	STAGES
2	1
2	2
4	3
3	4

	2	5

COST = 79.0000000000000

COST = 97.0000000000000

RELIABILITY = 0.0.86124487680

8.17

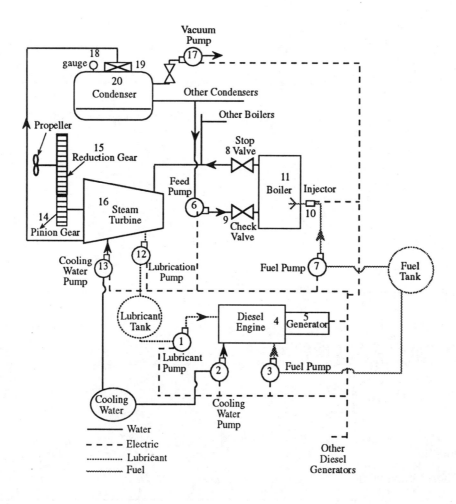

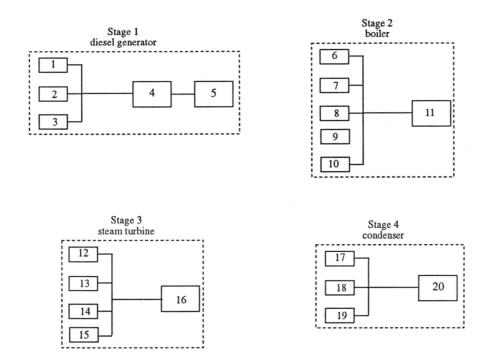

As shown, all the devices in one stage are serial connected to each other. If one of the component fail, the whole stage fails.

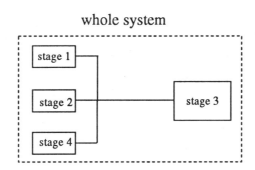

Note that all the stages are serial connected to each other. Any element fails, the whole system fails. Therefore, to increase the system reliability, more elements should be included in parrallel.

b) Optimize the system reliability for the cost limit of C_L $ 2,400.000. For simplicity the cost of the stages are divided by $10000.00. The output of the computer program given in Appendix D-2 is shown below

NUMBER OF ELEMENTS	STAGES
4	1
5	2
2	3
4	4

COST = 232.0000

RELIABILITY = 0.998162743604

$\Delta \lambda = 0.000005$ is used

The above results show that with the cost limit of $2,400,000, a ship will have 4 diesel generators, 5 boilers, 2 steam turbines, and 4 condensers.

c) Minimize the cost for a given reliability of at least 0.98. The output of the computer program given in Appendix D-3 is shown below

NUMBER OF ELEMENTS	STAGES
4	1
4	2
2	3
4	4

COST = 2,298,000

RELIABILITY = 0.9981086343243

From the above results, ship will have 4 diesel generators, 4 boilers, 2 steam turbines, and 4 condensers.

8.18

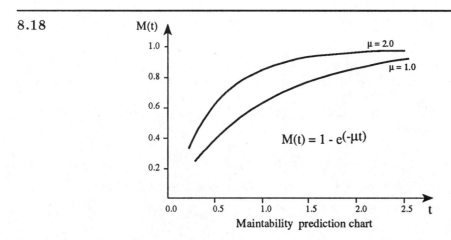

Maintability prediction chart

8.19

For Weibull distribution:

$$f(t) = \frac{bt^{b-1}}{\theta^b} exp[-(\frac{t}{\theta})^b] \qquad (8.19.1)$$

where, b is the shape factor, θ is the scale parameter, and t is the time.

$$M(t) = \int_0^t f(t)dt \qquad (8.19.2)$$

Substituting Equation (8.19.1) into (8.19.2) and integrating from 0 to t we have

$$M(t) = 1 - e^{(t/\theta)^b} \quad , \text{and}$$

$$MTTR = \theta \Gamma\left(\frac{b+1}{b}\right)$$

8.20

Steady state availability

$A = \frac{\mu}{\lambda+\mu} = \frac{0.8}{0.017+0.08} = 0.825$

Instantaneous availability

$A(t) = \frac{\mu}{\lambda+\mu} + \frac{\lambda}{\lambda+\mu} exp[-(\lambda+\mu)t] = \frac{0.08}{0.017+0.08} + \frac{0.017}{0.017+0.08} exp[-(0.017+0.08)20] = 0.834$

8.21

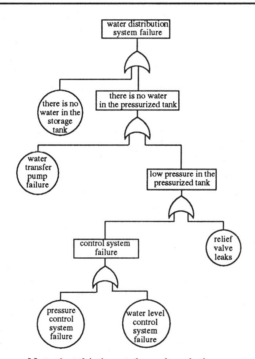

Note that this is not the only solution.

Chapter 10

Solutions for Case studies

NSPE Board of Ethical Review Case Discussions

NOTE:: The opinions included in this appendix are based on data submitted to the Board of Ethical Review and do not necessarily represent all of the pertinent facts when applied to a specific case. These opinions are for educational purposes only and should not be construed as expressing any opinion on the ethics of specific individuals.

10.1 Case Study 10.1

A) The engineers of Company "B" fulfilled their obligation under Section 1(c) of the Code by notifying their employer that they did not believe the project would be successful as designed by the engineers of Company "A". They also met the requirements of Section 2 in pointing out the consequences to be expected from proceeding under the original plans and specifications. By their actions the engineers of Company "B" regarded their "duty to the public welfare as paramount," as required by Section 2(a).

The further and more difficult question, however, is whether the engineers of Company "B" are required or ethically permitted to refuse to proceed with the production on the basis of plans and specifications which they continue to regard as unsafe.

In Case 61-10, we held that engineers assigned to the redesign of a commercial product of lower quality should not question the company's business decision, but had an obligation to point out any safety hazards in the new design. In that case, however, the redesign of the product involved only a question of a lower quality product and did not raise the problem of the product endangering public health or safety.

Section 2(c) of the Code is specific in holding that engineers will not complete, sign, or seal plans and/or specifications that are not of a design safe to the public health and welfare. In this situation the engineers of Company "B" have not been requested, or required, to "sign, or seal plans and/or specifications" at all. This has been done by the engineers of Company "A". A literal construction of the Code language may, therefore, indicate that the engineers of Company "B" may ethically proceed with their role in the production process. But we think that this is too narrow a reading of the Code and that the purpose and force of Section 2(c) is that the engineer will not participate in any way in engineering operations which endanger the public health and safety.

The last sentence of Section 2(c) is likewise clear in requiring that the engineers not only notify proper authority of the dangers which they believe to exist, but that they also "withdraw from further service on the project." This mandate applies to engineers serving clients or employers.

Where, as in this case, there is an apparent honest difference of opinion as to the safety features of the machinery between the engineers of Company "A" and the engineers of Company "B" it would be appropriate for the question to be referred to an impartial body of experts, such as a technical engineering society in the particular field of practice, for an independent determination.

So long as the engineers of Company "B" hold to their opinion that the machinery as originally designed and specified would be unsafe to the public they should refuse to participate in its processing or production under the mandate of Section 2(c). While such refusal to comply with the instruction of their employer may cause a most difficult situation, or even lead to the loss of employment, we must conclude that these considerations are subordinate to the requirements of the Code.

Conclusion:

The ethical obligations of the engineers of Company "B" are to notify their employer of possible dangers to the public safety and seek to have the design and specifications altered to make the machinery safe in their opinion; if the opinions cannot be reconciled they should propose submission of the problem to an independent and impartial body of experts: unless and until the engineers of Company "B" are satisfied that the machinery would not jeopardize the public safety they should refuse to participate in any engineering activity connected with the project.

B) In Case 65-12 we dealt with a situation in which a group of engineers believed that a product was unsafe, and we determined that so long as the engineers held to that view they were ethically justified in refusing to participate in the processing or production of the product in question. We recognized in that case that such action by the engineers would likely lead to loss of employment.

In Case 61-10 we distinguished a situation in which engineers had objected to the redesign of a commercial product, but which did not entail any question of public health or safety. On that basis we concluded that this was a business decision for management and did not entitle the engineers to question the decision on ethical grounds.

The Code section in point related to plans and specifications "that are not of a design safe to the public health and welfare," and ties that standard to the ethical duty of engineers to notify proper authority of the dangers and withdraw from further service on the project.

That is not quite the case before us; here the issue does not allege a danger to public health or safety, but is premised upon a claim of unsatisfactory plans and the unjustified expenditure of public funds. We could dismiss the case on the narrow ground that the Code does not apply to a claim not involving public health or safety, but we think that is too narrow a reading of the ethical duties of engineers engaged in activities having a substantial impact on defense expenditures or other substantial public expenditures that relate to "welfare" as set forth in Section III.2.b.

The situation presented here has become well known in recent years as "whistleblowing", and we note that there have been several cases evoking national interest in the defense field. As we recognized in earlier cases, if an engineer feels strongly that an employer's course of conduct is improper when related to public concerns, and if the engineer feels compelled to blow the whistle to expose the facts as he sees them, he may well have to pay the price of loss of employment. In some of the more notorious cases of recent years engineers have gone through such experiences and even if they have ultimately prevailed on legal or political grounds, the experience is not one to be undertaken lightly.

In this type of situation, we feel that the ethical duty or right of the engineer becomes a matter of personal conscience, but we are not willing to make a blanket statement that there is an ethical duty in these kinds of situations for the engineer to continue his campaign within the company, and make the issue one for public discussion. The Code only requires that the engineer withdraw from a project and report to proper authorities when the circumstances involve endangerment of the public health, safety, and welfare.

Conclusion:
Engineer A does not have an ethical obligation to continue his effort to secure a change in the policy of his employer under these circumstances, or to report his concerns to proper authority, but has an ethical right to do so as a matter of personal conscience.

10.2 Case Study 10.2

At first blush this appears to be a case involving a relatively small economic issue compared with the larger commercial and industrial projects with which engineers are often concerned. But as it involves an ethical principle we have not had occasion to address before, we will consider it on the broader philosophical aspects. Also, we note that this is not a case of an engineer allegedly violating the mandate of Section III.4. not to disclose confidential information concerning the business affairs of a client. That provision of the Code necessarily relates to confidential information given the engineer by the client in the course of providing services to the client. Here, however, there was not transmission of confidential information by the client to the engineer.

Whether or not the client in this case actually suffered an economic disadvantage by the reduction of its bargaining power in negotiating the price of the residence through the owner having knowledge gained from the inspection report, the same principle should apply in any case where the engineer voluntarily provides a copy of a report commissioned by a client to a party with an actual or potential adverse interest.

It is a common concept among engineers that their role is to be open and aboveboard and to deal in a straightforward way with the facts of a situation. This basic philosophy is found to a substantial degree throughout the Code, e.g., Sections II.3. and II.3.a. At the same time, Section II.1.c. recognizes the proprietary rights of cleints to have exclusive benefit of facts, data, and information obtained by the engineer on behalf of the client.

We read into this case an assumption that Engineer A acted without thought or consideration of any ulterior motive; that he, as a matter of course, considered it right and proper to make his findings known to all interested parties in order that the parties handle their negotiations for the property with both sides having the same factual data flowing from his services.

Thus, although we tend to exonerate Engineer A of substantial or deliberate wrongdoing, he was nevertheless incorrect in not recognizing the confidentiality of his relationship to the client. Even if the damage to the client, if any in fact, was slight, the principle of the right of confidentiality on behalf of the client prediminates.

Conclusion:
Engineer A acted unethically in submitting a copy of the home inspection to the real estate firm representing the owners.

10.3 Case Study 10.3

The facts in this case raise some of the most fundamental questions concerning the manner in which engineering firms engage in the practice of the profession. Probably the best starting point of this inquiry is an examination of the applicable Code provisions.

Sections II.2.a., II.2.b., and II.2.c. address many of the issues related to the facts of this case. Section II.2.a. seeks to admonish the engineer to accept work only in those areas of practice in which the engineer possesses the proper qualifications in order to competently perform the tasks to which he is assigned. Section II.2.b. examines the issue of ethical responsibility and states that an engineer must sign and seal documents and assume legal responsibility only for that in which he possesses understanding and cognizance. Finally, Section II.2.c. establishes a hieracrhy of responsibility by which engineers may coordinate and assume responsibility for entire projects as long as those individuals under the engineer's responsible control are identified as having prepared each technical segment of the work. The rationale behind those rules lies in the recognition that while the signature and seal of the engineer has consequences which go beyond the issue of ethics, the conduct of the engineer in the preparation of the plans and drawings involves the professional judgment and discretion of the engineer—judgment and discretion which are shaped by a variety of ethical concerns.

For example, in recent Case 85-3 where an engineer with experience and background solely in the field of chemical engineering accepted a position as a county surveyor, we noted that although the duties of the position included oversight of surveying reports and highway improvement but did not include actual preparation of engineering and surveying documents, nevertheless the engineer was unethical in accepting the position. As the Board noted: "It could be stated that Engineer A's responsibilities did not iclude actual preparation or approval of engineering or surveying documents, that instead such documents would be prepared or approved by qualified individuals, and that Engineer A's role would be to oversee those doucments and reports. We are convinced that neither is this the intent of the Code provisions nor is this what is commonly understood to be the proper oversight role of a county surveyor."

Clearly, in Case 85-3, the Board was faced with a situation in which an engineer was seeking to fulfill a role in which he possessed neither the qualifications nor the experience to perform in a competent manner. In the present case there is no indication that Engineer A possesses all of the qualifications or the experience to perform all of the requisite services. Rather, the issue here is the extent to which a professional engineer may ethically seal all of the documents the preparation of which he has delegated to subordinates.

Sections II.2.a., II.2.b., and II.2.c. are mutually dependent Code provisions which must be read together in order for them to have meaning. In the context of the instant case one of the most important aspects of the language of those provisions is the reference to "direction and control" found in Section II.2.b. We think a carefully crafted definition of that provision will assist us in a resolution of the facts in this case.

The term "direction" is generaly defined by *Webster's New Collegiate Dictionary* (1981 ed.) as "guidance or supervision of action or conduct; management; a channel or direct course of thought or action." The word "control" is generally defined as "the authority to guide or manage; direction, regulation, and coordination of business activites." It is clear that "direction" and "control" have a meaning which, when combined, would suggest that an engineer would be required to perform all tasks related to the preparation of the drawings, plans, and specifications in order for the engineer to ethically affix his seal. More pertinent to the engineering profession, however, is the National Council of Engineering Examiners Model Law, which is endorsed by this Board and reads as follows: responsible charge—the term "responsible charge" as used in this act shall mean "direct control and personal supervision of engineering work."

We recognize that the role of the chief engineer in an engineering firm may be that of a "manager who provides guidance, direction and counsel to those within his responsible charge." Indeed, in a large engineering firm this role is crucial to the successful operation of the firm. As in the facts in the instant case, the chief engineer should be involved at the outset of the project in the establishment of the design concept and the design requirements, as well as in the review of the various elements of the design or project status as the project develops. In addition, the chief engineer should be available to consult on technical questions relating to the project design.

To this end, we reiterate the language contained in Section II.2.c. noting that "each technical segment [shall be] signed and sealed only by the qualified engineers who prepared the segment." Finally, we would also note that whereas in the insant case the work is being performed by individials who are not licensed professional engineers, the firm has an ethical obligation that this work be performed under the direct control and personal supervision of registered engineers who would seal the document.

Conclusion:
It is unethical for Engineer A to seal plans that have not been prepared by him, or which he has not checked and reviewed in detail.

10.4 Case Study 10.4

The technical expertise that engineers can offer in the discussion of public issues is vital to the interests of the public. We have long encouraged engineers to become active and involved in matters concerning the well-being of the public. Moreover, the NSPE Code of Ethics makes clear that engineers should "seek opportunities to be of constructive service in civic affairs and work for the advancement of the safety, health and well-being of their community." (Section III.2.a.)

Obviously, this important involvement must be appropriate to the circumstance of the situation. In situations where an engineer is being asked to provide technical expertise to the public discussion, the engineer should offer objective, truthful, and dispassionate professional advice that is pertinent and relevant to the points at issue. The engineer should only render a professional opinion publicly, when that opinion is (1) based upon adequate knowledge of the facts and circumstance involved, and (2) the engineer clearly possesses the expertise to render such an opinion.

The Board has earlier visited situations in which engineers have publicly rendered professional opinions. In Case 65-9, a consulting engineer who had performed the engineering work on a portion of an interstate highway to which a proposed controversial highway by-pass would connect, issued a public letter which was published in the local press, criticizing the cost estimates of the engineers of the state highway department, stating alleged disadvantages of the proposed route, and pointing out an alternative route. The newspaper story contained the full text of the letter from the consulting engineer.

In deciding that it was ethical for the engineer to publicly express criticism of the proposed highway routes prepared by engineers of the state highway department, the Board stated: "...the whole purpose of engineering is to serve the public interest. When an engineering project has such a direct and substantial impact on the daily life of the citizenry as the location of a highway it is desirable that there be public discussion. The Code does not preclude engineers, as citizens, from participating in such public discussion. Those engineers who have a particular qualification in the field of engineering involved may be said to even have a responsibility to present public comment and suggestions in line with the philosophy expressed in the Code."

Thereafter, in Case 79-2, the Board ruled that where an engineer had significant environmental concerns, it was not unethical for the engineer to criticize a town engineer and a consulting engineer with respect to findings contained in a report on a sanitary landfill for the town. Said the Board: "It is axiomatic that an engineer's primary ethical responsibility is to follow the mandate of the Code to place the public welfare over all other considerations."

We noted that these issues in the public arena are subject to open public debate and resolution by appropriate public authority. Here the engineer was acting within the intent of the Code in raising his concern. We concluded by citing earlier decision Case 63-6 in which we noted: "There may also be honest differences of opinion among equally qualified engineers on the interpretation of the known physical facts...The Code does not prohibit...public criticism; it only requires that the engineer apply due restraint...in offering public criticism of the work of another engineer; the engineering witness will avoid personalities and abuse, and will base his criticism on the engineering conclusions or application of engineering data by offering alternative conclusions or analyses." It is clear, based upon the Code of Ethics and several interpretations of the Code by this Board that the engineer may and, indeed in some cases, must ethically provide technical judgment on a matter of public importance with the aforementioned considerations concerning expertise, adequacy of knowledge, and the avoidance of personality conflicts in mind.

However, we must note that under the facts of this case, we are not merely dealing with a disinterested engineer who on her own has decided to come forward and offer her professional views. Rather, we are dealing with an engineer who was retained by a newspaper to provide her professional opinion with the understanding that the opinion could serve as the basis for news articles concerning the safety of the bridge. This fact gives an added ethical dimension to the case and requires our additional analysis. In this regard, it is our view that as a condition of her retention by the newspaper involved, Engineer A has an ethical obligation to require that the newspaper clearly state in the articles that Engineer A had been retained for a fee by the newspaper in question to perform the one-day observation of the bridge site.

We should also add that in circumstances such as here where an engineer is being retained by a newspaper to offer a professional opinion concerning a matter of public concern, the engineer must act with particular care, should exercise the utmost integrity and dignity, and should take whatever reasonable steps are necessary to enhance the probability that the engineer's professional opinions are reported completely, accurately, and not out of context. While we recognize that there are limits to what an engineer can do in these areas, we believe that the engineer has an obligation to the public as well as to the profession to protect the integrity of her professional opinions and the manner in which these opinions are disseminated to the public.

Conclusion:

It was not unethical for Engineer A to agree to perform an investigation for the newspaper in the manner stated but Engineer A has an obligation to require the newspaper to state in the article that Engineer A had been retained for a fee by the newspaper to provide her professional opinion concerning the safety of the bridge.

10.5 Case Study 10.5

This Board has noted on numerous occasions that the ethical duty of the engineer in areas of conflict of interest is to inform the client of those business connections or interests that may influence the judgment and quality of the engineering services. Those decisions have been consistent with the provisions of Section II.4.a. of the NSPE Code of Ethics cited above.

While that provision of the Code has been interpreted many times over the years, it is, as are all Code provisions, subject to constant examination and reinterpretation. For any code of ethics to have meaning, it must be a living, breathing document which responds to situations that evolve and develop.

This Board has generally interpreted that Code provision in a strict manner. In BER Case 69-13, the Board reviewed a situation where an engineer was an officer in an incorporated engineering consulting firm that was engaged primarily in civil engineering projects for clients. Early in the engineer's life, he had acquired a tract of land by inheritance, which was in an area being developed for residential and industrial use. The engineer's firm had been retained to study and recommend a water and sewer system in the general area of his land interest. The question faced by the Board under those facts was, "May the engineer ethically design a water and sewer system in the general area of his land interest?" The Board ruled that the engineer could not ethically design the system under those circumstances.

The Board acknowledged that the question was a difficult one to resolve, pointing to the fact that there was no conflict of interest when the engineer entered his practice but that the conflict developed in the normal course of his practice when it became apparent that his study and recommendation could lead to the location of a water and sewer system that might cause a considerable appreciation in the value of his land depending upon the exact location of certain system elements in proximity to his land. The Board stated that while the engineer must make full disclosure of his personal interest to his client before proceeding with the project, such disclosure was not enough under the Code. Said the Board, "He can avoid such a conflict under these facts either by disposing of his land holding prior to undertaking the commission or by declining to perform the services if it is not feasible or desirable for him to dispose of his land at the particular time." The Board concluded by saying: "This is a harsh result, but so long as men are in their motivations somewhat 'lower than angels,' it is a necessary conclusion to achieve compliance with both the letter and the spirit of the Code of Ethics. The real test of ethical conduct is not when compliance with the Code comports with the interests of those it is intended to govern, but when compliance is adverse to personal interests."

We agree with much of what was stated in BER Case 69-13 considering the Code then in effect. In its reading of the Code of Ethics, the Board took a strict view of the meaning of the Code provisions then in force, which stated:

"8. The Engineer will endeavor to avoid a conflict of interest with his employer or client, but when unavoidable, the Engineer shall fully disclose the circumstances to his employer or client."

"8.(a) The Engineer will inform his client or employer of any business connections, interests, or circumstances which may be deemed as influencing his judgment or the quality of his services to his client or employer."

It is clear from a reading of BER Case 69-13 that the Board focused its attention on the first clause of Section 8 stating that "The engineer will endeavor to avoid a conflict of interest with his employer or client." Undoubtedly, the Board reasoned that this was the basic obligation of the engineer in this context, and that any qualification of that obligation would dilute the essential meaning and intent of that obligation. Therefore, the Board did not choose to rely upon the remaining provisions contained in Sections 8 and 8(a) in reaching its decision. Instead, the Board determined that under the facts it would not be sufficient for the engineer to make full disclosure of his personal interest to the client in order to properly address the potential conflict-of-interest question.

While the reasoning of the Board in BER Case 69-13 is extremely important in understanding the ethical dimensions of the instant case, the decision becomes less significant in view of the fact that the Code provisions under which the decision was rendered have been crucially altered. (See Code Sections II.4. and II.4.a., the successor provisions to Section 8.)

As one can readily see, the phrase "engineer will endeavor to avoid a conflict of interest with his employer or client..." is no longer contained in the applicable Code provision. Clearly, the reason for that omission is certainly not out of a lack of desire within the engineering profession for an ethical proscription relating to conflicts of interest. Truly, ethical dilemmas relating to conflicts of interest are some of the most significant issues facing the engineering profession today. Nevertheless, the provision in the Code relating to conflicts of interest was amended and those changes impact upon the manner in which this Board regards BER Case 69-13 as well as the manner in which the Board interprets the Code. It is evident that had Sections II.4. and II.4.a. been in effect at the time the Board decided BER Case 69-13, the Board may well have reached a different result.

While it is not our role to speculate upon the intent of this significant change in the NSPE Code of Ethics since BER Case 69-13 was rendered, we do think that some expression by this Board in that regard would assist readers in understanding the basis for the change. In no sense should this change be interpreted in any way to suggest a retreat by this Board or the Code of Ethics from a deep concern for dilemmas relating to conflicts of interest. Rather, it is our view that the modifications in the Code reflect recognition of the fact that conflicts of interest emerge in a multitude of degrees and circumstances and that a blanket, unqualified expression prohibiting engineers to avoid all activities that raise the shadow of a conflict of interest is not a workable approach.

It is often a question of degree as to what does and does not constitute a significant conflict of interest. Obvious and significant conflicts of interest are easily identifiable and should always be avoided. These difficult, multifaceted situations require discussion and consideration as they are complex and sometimes irresolvable. A code should address and provide guidance for these kinds of conflicts of interest. We believe the new Code provisions sought to establish the ethical obligation to engage in dialogue with a client or employer on the difficult questions relating to conflicts of interest. We think that it was for this reason that the Code provisions were altered.

Turning to the facts of the instant case, we are of the view that the ethical obligations contained in Section II.4.a. do not require the engineer to "avoid" any and all situations that may or may not raise the specter of a conflict of interest. Such an interpretation of the Code would leave engineers with neither any real understanding of the ethical issues nor any guidance as to how to deal with the problem. The basic purpose of a code of ethics is to provide the engineering profession with a better awareness and understanding of ethical issues that impact upon the public. Only through interacting with the public and clients will engineers be able to comprehend the true dimensions of ethical issues. We believe that holds true in the area of conflicts of interest.

We add that the Board assumes that under the facts of this case, the state agency involved has a fully qualified staff which will ultimately review the recommendation of the engineer.

Therefore, we are of the view that Engineer A's discussion with the client prior to performing the services and disclosing the possible conflict of interest came within the ethical guidelines of the Code and was a proper course to take in dealing with the conflict. We are not willing to state as we did in BER Case 69-13 that the engineer can only avoid such a conflict either by "disposing of his land and holdings prior to undertaking the commission or by declining to perform the services if it is not feasible or desirable for him to dispose of his land at the particular time." We do not read the current Code to require such action.

Conclusion:
It was not unethical for Engineer A to perform the feasibility study despite the fact that his land may be affected thereby.

10.6 Case Study 10.6

The facts of this case raise a number of issues pertinent to various provisions of the Code of Ethics. However, before this Board examines those several Code provisions, we think it would be appropriate to examine earlier BER decisions relating to the issues present here.

In BER Case 76-3, a decision involving an engineer appearing as an expert witness for a private development company before a county board while serving as a paid consultant to the county, this Board noted that the Code of Ethics requires an engineer to endeavor to avoid a conflict of interest: "When Engineer A was approached, while still on retainer to the county, by the development company, it should have been quite clear to him that a conflict of interest was inevitable."

It seemed in that case that a little interrogation of the development company concerning its plans would have revealed the conflict of interest. Under the facts of the case, Engineer A's role as an expert witness in the ordinary sense of that kind of professional arrangement would be unacceptable. Engineer A was doing more than offering his expertise in engineering matters as an aid to a fuller understanding by the county board; he was in fact a paid advocate of a private interest in open conflict with the engineering opinions of the county engineers.

In BER Case 76-3, this Board distinguished that case from earlier BER Case 74-2 in which the Board held that a part-time consultant arrangement to municipalities by engineers in private practice did not preclude those same engineers from providing normal engineering service to the same municipalities. We then noted that the key distinction between BER Case 74-2 and BER Case 76-3 was that in BER Case 74-2 the engineer's loyalties were not divided, whereas in BER Case 76-3 Engineer A is seen to be on both sides of the divided issue.

More recently in BER Case 82-2, a decision involving an engineer who prepared a home inspection report for a client, a potential home purchaser, and thereafter released the contents of the report to the real estate firm representing the seller of the home without the consent of the client, the Board ruled that this action was not in accord with the Code of Ethics. In that case this Board noted Section II.1.c.

During the same term, in BER Case 82-6, this Board ruled that where an engineer is retained by the U.S. government to study the causes of a dam failure, it would be unethical for the engineer to agree to be retained by the contractor involved in the construction of the dam. The contractor had filed a claim against the U.S. government for additional compensation. Citing the provisions of Section III.4.b., we found that there was nothing in the record to indicate that the engineer was given the consent of his former client, the U.S. government, to represent the interests of the contractor in its claim against the government for additional compensation.

All of the aforementioned cases represent longheld BER views relating to the question of conflicts of interest and the duty of engineers who gain, or are perceived to have gained, access to knowledge that may be advantageous to one client and disadvantageous to another. In some instances, it has been suggested by this Board that under certain circumstances, it may be appropriate for an engineer to first resign a particular position, such as consultant to a municipality, before agreeing to perform services for a client that might have a conflicting interest. (See BER Case 76-3.) Obviously, the degree to which this may be the proper, ethical course of action is dictated by the particular facts and circumstances of a case.

In the present case, the mere fact that Engineer A ceased performing services for Attorney Z would not be an adequate solution to the ethical dilemma at hand. Nor is the fact that Engineer A has agreed to provide a "separate and independent engineering and safety analysis report." On the former point, the fact that Engineer A ceased performing services for Attorney Z does not mitigate the fact that Engineer A throughout his first analysis had access to information, documents, etc., that were made available to him by the plaintiff and plaintiff's attorney in a cooperative and mutually beneficial manner. This Board cannot accept the proposition that following the termination of his relationship with attorney for plaintiff he would "blot all" of that information from his mind and start from "square one" in performing his engineering and safety analysis report. Nor do we believe the latter point that Engineer A would be capable of providing a "separate and independent" report for the defendant in this case. (See also Section II.4.b.) It is clear from the facts that the real reason for the defendant's attorney's hiring Engineer A was that he believed Engineer A would provide a report that would be favorable. Engineer A had to have been aware of the reasons why his services were being retained by virtue of the sequence of events. Even if Engineer A was so naive as to believe that Attorney X was unaware of the circumstances of his termination, we believe that this would not excuse his actions. Section III.4.b. is clear in this regard. At a bare minimum, Engineer A should have fully discussed the issue with Attorney Z.

It may be argued, as was stated in the earlier BER Case 74-2, that Engineer A's loyalties under these facts were not divided because he had terminated his relationship with plaintiff's attorney. However, we must recognize that while Engineer A may not currently have a professional relationship with a former client, he still has an ethical obligation to that client to protect certain confidential information and facts, as well as a certain duty of trust and loyalty. How long that duty of trust and loyalty must be maintained we are not prepared to state at this time. However, we are certainly willing to state that such a duty exists for the duration of one legal proceeding.

Finally, with regard to the duty of the engineer to be objective in his professional reports and statements (II.3.a.), we note that it has sometimes been suggested that engineers who act as paid expert witnesses have an inherent conflict between their duty to tell the truth and their obligation to perform their services consistent with the best interests of the client. We note that in this case, Engineer A developed an analysis report that was inconsistent with the legal interests of the client. Under the facts, Engineer A did not act in the role as a "hired gun," seeking to testify in favor of the client who was paying his fee. We make this point to underscore the importance of forensic engineers "calling them as they see them." Had Engineer A ceased his involvement in the case following the termination of his relationship with Attorney Z, he would have been acting in a wholly ethical manner. His ethical transgressions were a result of his subsequent involvement with Attorney X.

Conclusion:

It was unethical for Engineer A to agree to provide a separate engineering and safety analysis report.

10.7 Case Study 10.7

The previous cases dealing with gifts have been under what may be called "reverse facts", in that the gifts were given by the engineers to those in the position of being able to influence the award of contracts for engineering services.

In Case 79-8, an engineer in private practice gave a gift to an engineer in a public agency, and we held that it was unethical for the one engineer to make the gift, and unethical for the other engineer to accept it. But the primary thrust of the discussion revolved around the Code section barring the use of gifts by engineers to secure work. In that emphasis we referred to the criteria established in Case 60-9 on the issue of whether the gift was of a nature which might influence the award of work to engineers.

In Case 76-6 we dealt again with a gift by an engineer to an official of a foreign country, and also ruled that, notwithstanding the practice in the foreign country for officials to receive gifts from those who do business with the agency of the foreign country, it was unethical for the engineer to offer the gift.

In the first case involving gifts (Case 60-9), we looked at three circumstances, one of which involved engineering employees of an industrial company, who were in a position to recommend for or against the purchase of products, accepting nominal cash gifts from a salesman of particular products being offered their employer. In that case we also held that acceptance of even nominal gifts raised a question of integrity and hence was unethical.

The emphasis in the case before us raises more pertinently the idea of engineers accepting, rather than giving, gifts. Applying the principles of the earlier cases, and the language of the Code, as cited above in several sections, it seems clear that there would be, at least, a reasonable suspicion to others, and particularly to other contractors and suppliers, that acceptance of the gifts by Engineers A, B, and C would imply favoritism.

The language of Section II.4.c. covers acceptance of gifts, as well as their solicitation by engineers, and extends to the impact of such action on clients. Thus, the clients (developers) of the engineering firm may be led to question whether the recommendation of particular contractors or suppliers is totally unbiased and represents the independent judgment of the consulting firm. The first part of Section III.5.b. deals with political contributions, but applies equally to offering any gift in order to secure work. While under that language, Engineers A, B, and C did not in this case "offer" a gift, the section represents the same philosophy as Section II.4.c. And we have also cited Section III.5.b. , even though its emphasis is on acceptance of commissions or allowances from contractors dealing with clients of the engineer, because it is a reflection of the same ethical concept noted above.

When read in the context of the thrust of the entire Code on the matter of gifts, and in line with the ethical precept we have stated in the earlier cited cases, we affirm the overriding principle that engineers should "lean over backward" to avoid acceptance of gifts from those with whom they, or their firm, do business. We leave aside for this case the related issue of when a gift is permissible in terms of an expression of friendship or social custom, such as a calendar, a cigar, or picking up the check at a luncheon meeting. For general guidance on that point we refer the reader to the discussion in Case 60-9, namely that the guideline is that a gift of that nature be limited to those which will not raise any inference of compromising the independent professional judgment of the engineer, or that the giving or acceptance of such a gift be governed by the normal standards of good taste and acceptable custom.

Conclusion:

It was not ethical for Engineers A, B, and C to accept gifts from the contractors and suppliers.

10.8 Case Study 10.8

We have considered cases of this type on a number of occasions. In one, Case 62-7, an engineering consultant had been retained by a county commission to perform all necessary engineering and advisory services. The commission did not have an engineering staff so the engineer acted as the staff for the commission in the preparation of sewage and water studies, the financing of sanitary districts, and the approval of plans submitted by others. The engineer was also retained by a private company to perform engineering design for a development of several thousand housing units which involved extensive contract negotiations between the commission and the developer. We found that the engineer was in a position of passing engineering judgment on behalf of the commission on work or contract arrangements which the engineer performed or in which he participated. This obviously involved the self-interest of the engineer and divided his loyalties. Even if the engineer acted with the best of intentions, he was put into the position of assessing his recommendations to two clients with possibly opposing interests. Given these realities, we concluded that a conflict of interest existed.

More recently in Case 74-2, a case in which a state law required every municipality to retain a municipal engineer with that engineer's firm usually retained for engineering services for capital improvements needed by the municipality, we found that the engineer was not a bona fide "employee" of the municipality but a consultant, thus it was not unethical for him to serve as "municipal engineer" and participate in a consulting firm providing engineering services to the municipality. We reasoned that the public interest was best served by providing to small municipalities the most competent engineering services which they could acquire. It was assumed that the state law was intended to achieve that end.

In all honesty, it is difficult to reconcile these two cases, as the two cases were based in pertinent part on identical language. Both Case 62-7 and Case 74-2 were decided under the previous Code of Ethics, Section 8(b). That Code provision stated: "When in public service as a member, advisor, or employee of a governmental body or department, an engineer shall not participate in consideration of actions with respect to services provided by him or his organization in private engineering practice." (emphasis added)

In July, 1981, the Code of Ethics was revised and the substituted Section II.4.d replaced the above emphasized words with "...in decisions." We believe this change is significant and particularly relevant to this case.

Under the facts presented, Engineer A did not actually participate in "decisions" with respect to services solicited or provided by him or his organization in private or public engineering practice but rather reviewed, recommended, formulated, and oversaw plans. Although it is arguable that under the older Code provisions, Engineer A's activities would have constituted a conflict as he may have in fact participated in consideration of actions, it is our view that his activities were within the meaning of the amended Code provisions and did not constitute "decisions" under Section II.4.d. Therefore we conclude that one who serves as both city and county engineer for a retainer fee may provide private engineering consulting services to the city and county.

The question of whether an engineer who serves as a member of local boards or commissions which have some aspect of engineering may provide engineering services through his private firm to the boards and commissions was addressed in Case 75-7. We concluded there that an engineer serving on a commission could ethically provide services to the private owners because the engineer had abstained from the discussion and vote on permit applications. We cautioned, however, that care must be taken that the engineer in such a situation not have taken any action to influence the favorable decision on the permit. In this case, there is nothing to suggest Engineer A had taken any action to influence decisions as administrator of the city block grant program or as project administrator of the county airport authority. Therefore we find under the facts presented that Engineer A may properly provide engineering services through his private firm to the two city and county programs.

Finally, in Case 67-12, we indicated that when an engineer serves as a part-time county engineer and as a private consultant and in the latter capacity submits the plans of a private developer to the county for approval, he should not offer any recommendation for their approval. To do so is a useless act because it is basic to the Code that an engineer will not submit plans or other work which he does not believe represents the best interests of his client. Today we affirm that view. We cannot see how an engineer can wear two hats and still represent the best interest of his client. To do so would constitute a conflict of interest. If the county wishes to obtain a recommendation on the merit of his work, it is our view that it should retain another engineer for review in accordance with Section III.8.a. In our judgment it would be preferable for an engineer to avoid, entirely, situations as described in Question 3.

Conclusions:

Q1. It would be ethical for Engineer A, who serves as city engineer and county engineer for a retainer fee, to provide engineering services in a private capacity to the city or the county.

Q2. It would be ethical for Engineer A, who serves as a member of local boards or commissions which sometimes require the services of engineers, to provide services through his private firm to those boards and commissions.

Q3. It would be unethical for Engineer A, who serves as city engineer and county engineer for a retainer fee, to provide approval or render judgment on behalf of the city and/or county relative to projects on which Engineer A has furnished services through a private client.

10.9 Case Study 10.9

The case presented before the Board raises a number of significant points that have heretofore not been specifically addressed. In BER Case 77-11, the Board ruled that four engineers who founded a new firm did not violate the Code of Ethics by generally seeking work from former clients of their previous employer, but were in violation of the Code with regard to projects for which they had particular knowledge while working for their former employer. Although at first glance the facts in Case 77-11 appear to be quite similar to the instant case, they are distinguishable on two very important points: (1) In the instant case Engineer A notified "current" and not former clients of Engineer B and offered professional services to them. (2) Engineer A was still employed by Engineer B when Engineer A notified the clients and others of the offer of professional services. We are therefore now asked to decide whether one engineer in the employ of another who is aware of a pending termination may ethically contact "current" clients of an employer and offer professional services to the client without informing the employer.

An engineer is expected to act, at all times in professional matters for the employer, as a faithful agent and trustee (Section I.4). That requires the engineer to recognize both a duty of loyalty and good faith. An essential aspect of those is the duty to disclose. Certainly it is not possible for an engineer to meet those obligations to the employer if the engineer is engaging in such promotional activity to the employer's detriment We do not mean to suggest that an employee who severs all ties with the employer and then seeks to contact clients of the employer in order to offer engineering services is in violation of the Code. To the contrary, those were the facts of Case 77-11 and that case remains a proper interpretation of the Code. Nor do we wish to suggest any restraint exists upon one's absolute right to select in all cases, the engineer of one's choice. As we noted in Case 77-11, "We have often held that (the Code) is not to be interpreted to give an engineer or firm a right to prevent other engineers from attempting to serve former clients of other firms." Nevertheless, for the above-noted reason, it is concluded that Engineer A violated Section I.4. by failing to act as a faithful employee.

Another issue related to the conduct of Engineer A is whether Engineer A violated Section III.7. by competing with Engineer B using "questionable methods". It seems obvious that by failing to act as a faithful employee and by failing to disclose the actions to Engineer B, Engineer A engaged in questionable methods of competition. Even if Engineer A was not certain that the actions constituted unethical conduct, Engineer A knew or should have known that they were problematic and dubious and raised the possibility of an ethical violation. Therefore, we are of the view that Engineer A was in violation of Section III.7.

A related question under the facts of this case is whether Engineer A violated a duty of disclosure to all interested parties by entering into promotional efforts for work as a principal in connection with work for which Engineer A had gained a particular and specialized knowledge. The facts do not indicate whether Engineer A was attempting to secure work through particular and specialized knowledge gained. Assuming that in fact Engineer A had gained such knowledge and then sought such work without full disclosure to the employer, Engineer B, it appears that Engineer A would have violated Section III.4.a. of the Code. Again, Engineer A owes duties of loyalty, good faith, and disclosure to the employer for which the breach constitutes a violation of the Code. As an employee of Engineer B, engineer A could not ethically use proprietary information concerning clients, trade secrets, or other valuable information of the employer without full disclosure to the employer.

The other dimension of this case is the actions of Engineer B. Section II.5.a. of the Code specifically states that brochures or other presentations incident to the solicitation of employment shall not misrepresent pertinent facts concerning employers, employees, associates . . . with the intent and purpose of enhancing their qualifications and their work. Thus, the Code provision requires the Board to interpret that provision to determine whether (1) Engineer B in fact misrepresented "pertinent facts" and (2) whether it was the intent and purpose of Engineer B to "enhance the firm's qualifications and work." Both prongs must be present for a violation of Section II.5.a. to exist.

"Pertinent facts" are those facts that have a clear and decisive relevance to a matter at hand. Another way to characterize pertinent facts is as those that are "relevant and highly significant." It is not unusual for an engineering firm that seeks to promote itself for business reasons to include in such a brochure a statement of the firm's experience, its history, its qualifications, and the names and qualifications of the members of the firm. The names of the firm's members are often quite significant to the client selecting the firm. The client may be familiar with an individual member of the firm and the selection of that firm may be based on the presence of that engineer in the firm as represented in the brochure. It is clear, therefore, that the inclusion of the name of Engineer A in the firm's brochure constituted a misrepresentation of "pertinent facts."

The second point of inquiry is whether it was the "intent and purpose" of Engineer B to "enhance the firm's qualifications and work" by including Engineer A's name in the promotional brochure after Engineer A had left the firm. The facts presented in the case appear to demonstrate that Engineer B acted with "intent and purpose" in distributing the misleading brochure. Certainly, Engineer B was well aware of the impending termination of Engineer A. Engineer B was the very person who terminated Engineer A. Engineer B distributed the brochure while Engineer A was still employed but had been given a notice of termination by Engineer B. That could easily mislead potential clients into believing that Engineer A, noted as a key employee, would be available in the firm for consultation on future projects. Moreover, Engineer B distributed the brochure after Engineer A had left the firm. That is a clear misrepresentation of a pertinent fact with the intent to enhance the firm's qualifications and as such constitutes a violation of the Code.

Section III.3.a. states in part that "Engineers shall avoid use of statements containing a material misrepresentation of fact or omitting a material fact necessary to keep statements from being misleading; statements intended or likely to create an unjustified expectation. ..." Although that section appears to provide Engineer B with the appropriate guidance under the facts in this case, we are of the view that a requirement that Engineer B insert an addendum or an amendment in the brochure informing prospective clients that Engineer A would soon be leaving the firm is both impracticable and unnecessary. That would be a burden to all firms from the standpoint of both time and cost. We do believe that during the interim period between Engineer A's being given notice of termination and his actual cessation of employment, Engineer B had an obligation, during negotiations with a prospective client, to inform the client of Engineer A's pending termination. However, once Engineer A had been formally dismissed, Engineer B had an ethical obligation to cease using the brochure with Engineer A's name in it entirely.

Conclusions:

Q1. It was unethical for Engineer A to notify clients of Engineer B that Engineer A was planning to start a firm and would appreciate being considered for work while still in the employ of Engineer B.

Q2. It was not unethical for Engineer B to distribute a previously printed brochure listing Engineer A as a key employee provided Engineer B apprised the prospective client during the negotiation of Engineer A's pending termination.

Q3. It was unethical for Engineer B to distribute a brochure listing Engineer A as a key employee after Engineer A's actual termination.

10.10 Case Study 10.10

A) We have treated various aspects of political activity and political contributions by engineers under a variety of circumstances, the most recent involving the question of the amounts which engineers engaged in public project work may contribute through a political action committee (Case 75-13). In two earlier cases we dealt also with the amounts an engineer may contribute as related to securing engineering assignments (Case 60-9 and Case 73-6). Other political contribution or political influence cases have dealt with political influence not involving political contributions (Case 69-1) and assistance from a Congressman to secure a contract (Case 66-6).

We have cited these earlier cases to indicate the broad range of issues which may arise for engineers performing or seeking to secure contracts for their services from public bodies, which by definition are political bodies. A principal theme throughout these cases has been the recognition of dual considerations inherent in the political climate by which we govern ourselves: the right (some would say duty) for engineers along with all other citizens to actively participate ;in the political process and play a role in influencing the selection of public officials and at the same time the ethical restriction engineers have placed on themselves not to utilize political influence to secure engineering contracts.

It is basic to the profession's ethical standards that engineers be selected for engagements only on the basis of merit and qualification. Thus, the profession has historically gone to great lengths to spell out in precise detail the proper procedures for the selection of engineers by both public and private clients on the sole basis of qualification. Yet it would be naive not to recognize that in the case of public bodies the choice of the best qualified engineer or firm may be influenced in the minds of the public authorities by elements of personal relationship, associated civic activities, reputation in the community, and even political considerations. Although the selection should be made objectively to the extent possible, in the final analysis the choice of best qualified must be made by humans who are subject to the kinds of intangible relationships and backgrounds noted above.

Trying to balance these conflicting factors both in the code and in actual practical operations, the best the profession can do is evaluate specific factual circumstances in which a political relationship may exist. Thus we have attempted previously to identify guidelines as to the amount of political contribution, or express an adverse view on some forms of direct political influence.

As set forth in Case 73-6, the balancing test in trying to define these "thin line" relationships is "how the profession and the public may properly or improperly relate the amount of a political contribution to the later receipt of contracts from the recipient of the earlier contribution, directly or indirectly."

In the facts before us, the "contribution" is not directly financial, but realistically the activity of Roe has direct financial implications, and in a real sense his effort to raise "substantial" funds for the candidates will be regarded by the public in the same light as those who directly give financial support.

We also recognized this aspect of the problem in Case 73-6 in the comment that "to actively participate in the political process cannot be construed to mean solely the donation of money to the coffers of a candidate or party." And further in that case we applied the "Caesar's wife" test, first enunciated in Case 62-12.

Applying that same kind of rationale and logic to the case before us, we further note the application of the principles of section 1(g), section 3, and section 4 of the code, all of which in somewhat different words establish the basic premise that in case of doubt the engineer must refrain from conduct which may create an unfavorable impression on the part of the public. How far that principle may extend in the political arena or other differing circumstances we leave to future factual situations. But in the facts before us we have little doubt that the connection between the fund raising activities of Roe and his continuing acceptance of contracts from those he supported by his activity must lead to public opinion conclusions that there is a clear degree of self interest, and strong suspicion that Roe intends to obtain a direct financial benefit from his political involvement. Under that circumstance it follows that there will be an adverse public reflection on the dignity and honor of the profession.

We would emphasize, however, that what we have said in this case should not be construed as a blanket condemnation of political activity by engineers, a practice which should be continued and expanded. Rather, the line we have drawn should indicate to the profession that in undertaking political efforts the engineer who is in or may be in a position to secure advantage from that activity must make what may be a hard choice—refrain from such political involvement or remove himself from consideration for assignments or other benefits which may be construed to flow from his political support.

Conclusion:

It was not ethical for Roe to continue to accept engineering assignments from the county board having engaged in raising political funds for incumbent members of the county board.

B) The Board had previously decided a case with facts similar to those presented in the instant case. In an earlier decision, Case 62-4, the Board held that it was permissible for an engineering firm to "employ" a non-engineer as a representative of the firm to solicit work, provided that the non-engineer representative did not discuss engineering aspects of a project (including contract negotiations) with a prospective client. The Board noted in that decision that while it was not unethical to compensate the employed representative on a commission basis, that method of competition was undesirable since it could lead to a loss of confidence by the public in the professional nature of engineering services. Later, in Case 77-1 the Board found an ethical violation existed where an engineering firm paid a commission to a commercial marketing firm to secure work for it. The Board distinguished the two cases on the ground that in the former case, the firm had actually "hired" an employee to perform the solicitation function while in the latter case the firm retained an independent, outside firm for marketing purposes. Said the Board: ". . . the important difference to note in the facts before us is that the engineering firm has control over the conduct of an employee, whereas it has little or no control over the conduct of an outside marketing firm which operates on a commercial basis. The danger is thus much enhanced that a commercial marketing firm may more readily in its zeal to earn its compensation engage in conduct which may adversely reflect upon the dignity or honor of the profession."

A year later, in Case 78-7, the Board, faced with a similar set of facts, ruled that an engineering firm may not ethically enter into a marketing agreement with an individual and independent professional engineer on a commission basis. The Board, reviewing both Case 62-4 and Case 77-1, noted then that the Code of Ethics, as amended, contained a prohibition against the payment of commissions of any kind and that the Board was compelled to find that arrangement was impermissible under the Code.

Since the decision in Case 78-7 was rendered, the NSPE Code of Ethics has again been amended and now contains an exception to its prohibition against the payment of commissions in securing work. That exception appears to allow the payment of commissions in order to secure work to a "bona fide employee," or "bona fide established commercial or marketing agency" retained by an engineering firm. (See Section II.5.b.)

The only question presented to the Board in the instant case is whether the rules as established in earlier BER cases and in the present Code of Ethics would permit an engineer to enter into agreement whereby an independent landscape architect could refer clients to the engineer in return for a fee over and above the actual value of services rendered.

It is clear that a firm is no longer required, as was the case in Case 62-4 to "hire" a marketing representative as a member of its staff in order for its actions to come within the Code. It is equally clear that if Cases 77-1 and 78-7 were being decided today the results might be different in view of the fact that the Code now permits the payment of commissions to "bona fide commercial or marketing agencies." However, the facts in the instant case are a good deal different from both Case 77-1 and Case 78-7. Unlike Case 77-1 and Case 78-7, there is nothing in this case to indicate that the landscape architect is a "bona fide marketing agency." To the contrary, it appears that the landscape architect is wearing at least two hats and is wearing those hats simultaneously. The landscape architect proposed to act both as a marketing representative for Engineer A and, at the same time, expected to perform services at an inflated rate in connection with the work that the landscape architect secured for Engineer A. Such conduct does not demonstrate the requisite good faith, integrity of dealing, and honesty implicit in the definition of a "bona fide marketing agency," as required by Sections II.4. and II.5. of the Code.

Conclusion:

It was unethical for Engineer A to accept the proposal by the landscape architect to refer clients to Engineer A in return for a fee over and above the value of the landscape work which the landscape architect would presumably perform on each of the projects.

10.11 Case Study 10.11

The facts of the case presented to the Board, at first glance, appear to be fairly straightforward and easily addressed by the Code of Ethics. On its face we are presented with an engineer who has been retained by a client to design a project. However, both parties cannot agree as to the ultimate success of the project as developed by Engineer A. Thus, the client seeks to terminate the services of Engineer A, but wishes to obtain the drawings, plans, and specifications from Engineer A for a fee. Our discussion will be limited to the ethical rather than the contractual considerations of this case.

Much of the language contained in the Code relates to the engineer's obligation to protect public health, property, and welfare (Section II.1.a.). In the present case it appears that Engineer A had a strong concern for the protection of the public health and welfare. Nevertheless, it is the view of this Board that Engineer A could have delivered over the drawings to the client and his conduct would have been ethically proper.

While it is true that Engineer A has an ethical obligation under Section II.1.a., that obligation assumes that Engineer A is in possession of verifiable facts or evidence which would substantiate a charge that an actual danger to the public health or safety exists. In the instant case, Engineer A makes the overly broad assumption that if he were to deliver over to the client the drawings so that the client can present them to Engineer B to assist Engineer B in completing the project to the client's liking, Engineer B would develop a set of plans which would endanger the public health and safety. We think that such an assumption is ill-founded and is not based upon anything more than a supposition by Engineer A. Therefore, we are of the view that Engineer A should not have withheld the drawings on the basis of Section II.1.a.

In reviewing the conduct of Engineer A up until his refusal to deliver over the drawings to the client, we are of the view that Engineer A went as far as he was ethically required to go in preparing what he believed was the best design for the project and in informing the client of the dangers of proceeding with the client's simplified solution. Section III.1.b. is very clear in stating an "Engineer shall advise [his] client . . . when [he] believes a project will not be successful." We are of the view that, by conferring with the client and explaining his concerns over a proposed simplified solution, Engineer A had met his ethical responsibility.

In the event, however, that Engineer A does deliver over to the client the plans so that the client can present them to Engineer B for completion of the project to the client's liking, and thereafter Engineer A discovers that Engineer B developed plans which constitute a danger to the public, certain actions would then be required by Engineer A under the Code. Any verifiable conduct on the part of Engineer B which indicates that Engineer B's plans are a danger to the public, should be brought to the attention of the proper authorities, i.e., the responsible professional societies or the state engineering registration board.

Conclusion:

It would be ethical under the above circumstances for Engineer A to deliver over the plans and specifications to the client.

10.12 Case Study 10.12

As has been frequently stated by this Board and is clearly stated in the NSPE Code of Ethics, engineers are encouraged to participate in civic affairs and to become involved in political activity. This position is embodied in Section III.2.a. As has been noted before, this provision is a recognition of the valuable and unique perspective of the engineer and the enormous contribution that the engineer can make to public policy debates.

Certainly, participation by the engineer in the sphere of public policy must be tempered by a sense of reason and rationality. Engineers are expected to act in such matters in a responsible and prudent manner. While no one would ever suggest that engineers should not be opinionated or even vigorous in their political views, we think that it is correct to state that engineers have an ethical obligation to conduct such activities with an eye on objectivity and truthfulness. Without these basic guidelines, the engineer is in danger of losing credibility among members of both profession and the community as a whole.

Under the facts of this case there appears to be a genuine question as to whether Engineer A's actions were in an objective and truthful manner as required by Section II.3. The most obvious point seems to be that the comments were made primarily for political purposes—to drum up support among union employees by suggesting that Engineer A is sympathetic to their cause. The action also appears to have been made to provide Engineer A with a great deal of media exposure before the television cameras.

While it is certainly arguable that Engineer A was legitimately concerned with the issues of unsafe working conditions at the plant and what he saw to be management indifference, another issue of concern is the manner in which Engineer A addressed the issues of unsafe working conditions and management indifference. Rather than examining the allegation and attempting to mediate the differences between the parties, Engineer A appears to have furthered the conflict by making rhetorical pronouncements. By holding a placard that accused the company of "callous indifference" to the workers, Engineer A injected himself into the controversy and lost any and all appearances of impartiality. Engineer A attempted to exploit an extremely unpleasant situation for political gain.

Finally, the Board is concerned with the actions of Engineer A because it appears that Engineer A was promoting his own interest at the expense of the dignity and integrity of the profession (Section III.1.f.). Under the facts, there is little doubt that Engineer A's act of thrusting himself before television cameras with the placard in hand, without thoroughly investigating the specific conditions within the plant, suggests that Engineer A was seeking to promote his own interests, i.e., his political career, at the expense and dignity of the profession.

There appears to be some question here as to whether Section III.1.e. applies to Engineer A. That provision related to individuals who, as employees, actively participate in strikes, picket lines, or other collective coercive action. In this case, Engineer A was not an employee but was a candidate for office who was sympathetic to a particular cause. While we do not condone Engineer A's actions, we do not think Section III.1.e. is applicable to this case.

Conclusion:
It was unethical for Engineer A to accuse the company of callous disregard for the workers at the plant.

10.13 Case Study 10.13

We have cited an extraordinary number of pertinent Code references in this case, perhaps because of the uniqueness of the facts, and to bring out a central point of the Code–the duty of the engineer to act within stated ethical bounds in relationships with employers, as well as with clients.

While we tend to think of engineers in private practice in relation to clients, there is an equally important relationship between the engineer and his or her employer. It is easy to overlook the fact that even in consulting practice engineers tend to be employees of their firms, even if they are sole owners or principals of high rank in the firm.

We need not, we believe, in this set of circumstances, detail the application of the many cited sections of the Code. Starting with the mandate that engineers shall act in professional matters for each employer as a "faithful agent", it seems beyond peradventure that any case can be made for the conduct of Engineer A toward his employer. Here was a blatant effort to use the professional employee-employer relationship as a means for personal advancement in direct conflict with the interests of the employer.

Engineer A may contend that what he did was a business affair, rather than one involving ethical obligations. We must reject that distinction because almost every unethical action stipulated in the Code can be argued as one involving business interests. Perhaps the main lesson and point of this case is that the ethical and business activities of engineers cannot be simply or totally divorced; however labeled by the engineer, his or her conduct must be judged under the strictures of the code.

In addition to the primary duty of the engineer to be a faithful agent, as noted above, the Code demands integrity, which was flouted in this case. It likewise imposes a duty to subordinate private interest when that interest would conflict with the interest of the employer. There may be other sets of facts that would permit–or even require–an engineer to act contrary to the interests of an employer when there is an overriding public interest, but there can be no question in this case that Engineer A has offended the rule against unfair methods to seek advancement through taking advantage of his position in the firm; all leading to the clear that he acted by improper or questionable methods.

Conclusion:
Engineer A was unethical in making his demands under the stated circumstances.

10.14 Case Study 10.14

The "Rent-an-Engineer Plan" appears on the surface to be more a "job shop" operation than one that would be conducted by a firm engaged in normal engineering and construction operations. As set forth in BER Case 83-4, job shops are essentially employment referral agencies which list engineers and other technical personnel for temporary employment by others, taking a fee for their services. Such operations are well known in the technical world and are often employed by companies in need of temporary technical help.

In this case we are not concerned with the pros or cons of job shops, as such. Rather, we are called upon to evaluate the wording and tenor of the advertisement in light of the restrictions on advertising in Section II.3.a. of the Code of Ethics. For the purposes of this case, we assume that Smith & Jones, Inc., is owned or operated by engineers who are subject to the Code.

Both in language and in spirit the advertisement is ethically offensive. By reference to the "renting" of engineers it treats engineers as if they were the same as some kind of machinery or an inanimate object.

While the Code restrictions on advertising have been loosened to a degree in recent years on the basis of Supreme Court decisions barring a total ban on advertising of professional services, some restrictions are still allowed as reflected in the present Code language. Recently, the Chief Justice of the United States, speaking to the spread of advertising for legal services, commented on "the novel spectacle of lawyers advertising in newspapers, on radio, and on television in much the way that automobiles, dog food, cosmetics, and hair tonic are touted." The Chief Justice added that while the Supreme Court has said that attorneys (and by implication all of the professions) have a First Amendment right to advertise, "The very nature of a profession, as distinguished from the marketplace, is that standards of professional conduct are proscribed to protect the public. The standards have grown up slowly and painfully for centures. Codes of professional conduct take for granted that lawyers will not exercise every Constitutional right to its outer limits if to do so conflicts with higher professional standards."

We think that the statement of the Chief Justice exactly fits this case. Taken in the light of the wording of the Code of Ethics, the advertisement is intended to attract business by the use of showmanship and by use of what amounts to slogans and sensational language. It reflects adversely on the status and standing of the engineering profession by treating the members of that profession as a commodity and to that extent is conduct likely to discredit the profession under Section III.3. It is an example of "showmanship" for self-interest.

Aside from the merits of offering to "sell" the services of its engineering staff to others, there is no reason that Smith & Jones, Inc., could not make its offer on a dignified and ethically acceptable basis by a less flamboyant and offensive style of presentation.

Conclusion:
The advertisement of Smith & Jones, Inc., is a violation of the Code of Ethics.

10.15 Case Study 10.15

This case presents three distinct issues which, although not directly addressed by the Code of Ethics nor earlier BER decisions, are extremely important in regard to the integrity and honesty of intellectual work performed by university engineering faculty.

The first issue relates to that of engineering faculty using material from previous work performed and modifying that material in order to satisfy a requirement to publish. This development has occurred in recent years as a result of the emphasis placed by various universities and colleges upon the importance of publication. With pressures being exerted upon faculty to write articles acceptable for publication, some faculty, as a result of time pressures and other factors, have sometimes "cut corners" in order to satisfy the requirement to publish.

While we stress the importance of performing new and innovative engineering research, we are not convinced that previous work of a high quality could not form the basis of updated research by engineering faculty. Quite often engineering students and faculty embark upon areas of research, and owing to a variety of factors, many beyond their control (time constraints, priorities, funding, etc.), make the decision to postpone the reseasrch being conducted. Later, for a number of reasons, they may decide to resume the research. Flowing out of the concluded research may be articles or reports suitable for publication in technical journals. As long as an article is properly updated and the data verified and scrutinized in view of the time lapse, we are of the view that such publication would be entirely proper and ethical.

It may be suggested that because the earlier research was performed not as a faculty member but as an engineering student, the research was performed outside the scope of the faculty member's current employment and therefore should not be credited as research performed as faculty for the purpose of tenure. We have trouble accepting such an inflexible view, particularly in view of the aforementioned variables that may impact upon the ability to perform research. We think the better course to take is to examine the relative quality of the individual's research rather than to question the chronology of the research. As long as the research is of a high-quality nature, we are satisfied that no ethical violation exists. In view of the fact that the article was brought up to date and was ultimately published in a refereed journal, we are convinced that no ethical problem has emerged.

Turning to a second issue in this case, as noted earlier, we are sensitive to the extremely difficult position in which many faculty members have been placed with regard to the so-called rule of "publish or perish". This Board finds it extremely difficult to sanction a situation whereby Engineer A permits Engineer B, for whatever reason, to share joint authorship on an article when it is clear that Engineer B's contributions to the article are minimal. We think that Section III.3.c. speaks to this point. This Board cannot excuse the conduct of a faculty member who "takes the easy way out" and seeks credit for an article that he did not author. The only way a faculty tenure committee can effectively evaluate tenure candidates is to examine the candidates' qualifications and not the qualifications of someone else. For this Board to decide otherwise would be to sanction a practice entirely at odds with academic honesty and professional integrity. (See Section III.1.)

Finally, the facts of the case raise the question of Engineer A's ethical conduct in agreeing to include Engineer B as coauthor of the article as a favor in order to enhance Engineer B's chances of obtaining tenure. However genuine Engineer A's motives may have been under the circumstances, we unqualifiedly reject the action of Engineer A. By permitting Engineer B to misrepresent his achievements in this way, Engineer A has compromised his honesty and forfeited his integrity. Engineer A is unquestionably diminished by this action.

While this Board is fervent in its view and wishes to stress the importance of those three points, we also feel compelled to acknowledge that certain "gray areas" do exist. Frequently, technical articles are written that contain the names of many authors or contributors. Often it is difficult to identify in an objective manner the qualitative contributions of the various authors identified. While we recognize that this practice is a proper means of accurately identifying actual authors contributing to an article, we tend to be somewhat skeptical in general of this practice. We recognize the importance of collaboration in academic endeavors; however, we think that the collaborative effort should produce and reflect a high-quality product worthy of joint authorship, and should not merely be a means by which engineering faculty expand their list of achievements.

Conclusions:

Q1. It was ethical for Engineer A to use a paper he developed at an earlier time as the basis for an updated article.

Q2. It was unethical for Engineer B to accept credit for development of the article.

Q3. It was unethical for Engineer A to include Engineer B as coauthor of the article.

10.16   Case Study 10.16

Over the years the Board has been faced with a number of cases with facts similar to those present in the instant case. In Case 77-11, the Board ruled that four engineers who left the employ of a firm and founded a new firm, contracting the clients of their former firm, did not violate the Code of Ethics.

However, we found that the four were in violation of the Code with regard to projects for which they had particular knowledge while in the employ of the firm. As we understood the facts then, the four engineers did not undertake the promotional efforts with the former clients of the firm while in the employ of the firm, nor did they engage in negotiations for work while in the employ of the firm. We assumed that the four engineers possibly discussed among themselves the idea of soliciting work from former clients of the firm while still in its employ, but under a literal reading of that part of the Code, that degree of activity did not constitute a violation of the Code. However, we noted in Case 77-11 that such was not nearly so clear with regard to the latter portion of then-Section 7(a) (now Section III.4.a.) as related to practice in connection with a specific and specialized knowledge. We therefore found a violation of the Code provision.

Similarly in Case 79-10 where an engineer employed by a firm which was winding down its operations sought to offer his services to complete projects under his own responsibility and risk without the concurrence of the principal of his employing firm, we ruled that such a course of action would not be unethical. We did not construe the broader aspect of then-Section 7(a) as barring Engineer A from taking the step he contemplated if the client should agree to it. We noted that the thrust of Section 7(a) was that an engineer would not initiate self-promotional efforts to take over projects in which he has been involved while in the employ of another. In Case 79-10, however, the engineer did not initiate the idea of taking over the work of the projects; the idea was set in motion by his employer's proposed action. Accordingly, we found that the engineer would not be considered as attempting to compete unfairly under the principles embodied in the Code.

Reading both Case 77-11 and Case 79-10 suggests a need to balance (1) the interests of the firm and its interest in maintaining business goodwill with its clients, (2) the interests of the individually employed engineers, and most importantly (3) the interests of the client in retaining the firm of his choice. Obviously, the last interest is the most overriding interest of all. (See Code Section II.4.) No one can deny that a client has a right to retain the engineering firm of his choice. What must be addressed, however, is a method of effectuating that right in a manner which is both fair and equitable to all of the concerned parties.

We are of the view that whereas here a client approaches employed engineers who have been performing services for that client and initiates a request for services from those engineers independent of the firm, it would not be unethical for those engineers to agree to contract for professional services. We reach that conclusion here for two basic reasons. First, there does not appear to be any indication under the facts that Engineers X, Y, and Z had gained any "particular and specialized knowledge" which is required for a violation of Section III.4.a. Were we to interpret the term "particular and specialized knowledge" to embrace, for example, any information relating to the work, we would be compelled to find the conduct of Engineers X, Y, and Z to be unethical. However, we have chosen not to reach that finding inasmuch as most proposals are general in nature.

On our second point, this case, like many before, involves a variety of considerations which must be balanced. In this context we note that the facts here presented involve a client who has made a determination that he would like to proceed in a professional relationship with three engineers of his choice. It cannot be disputed that the essential purpose of the Code of Ethics is to protect the interests of the public and in particular the interests of clients in their relations with members of the profession. When faced with a situation where the Code of Ethics can be interpreted to protect either the interest of engineers or the interests of the public or the client, barring unforeseen circumstances, we are compelled to find in favor of the latter. For us to hold otherwise would be to obstruct the ability of an individual client to select those engineers of his choosing to perform professional services on his behalf.

Although not indicated by the facts, we can foresee a situation where a client would not want to continue a relationship with an engineering firm but would like to retain the services of individual employees of a firm and seeks to retain their services, we do not think the Code of Ethics should be interpreted to deny the right of a client to undertake such action.

Conclusion:
A strict interpretation of the Code under the facts of this case leads us to conclude that it would not be unethical for Engineers X, Y, and Z to agree to a contract for consulting services independent of Engineers A's firm.

10.17 Case Study 10.17

The ethical restriction cited has been in the Code of Ethics for a long period of time, but we have not heretofore had occasion to interpret it.

The word "engineers" is understood in the context of the full Code to refer to all engineers, whether as employers or employees. But to reflect the substantive purpose of the section, engineers in this context must refer only to those acting on behalf of an employer, since an employed engineer would not be accepting remuneration from another employee. However, while we have no question that this is the proper understanding and purpose of the section, we believe that consideration should be given to rewording to better reflect this intent.

The main purpose of the prohibition would appear to be grounded in the concept that employers should not require kickbacks from employees as a condition of employment. The salary offered should be the full salary without rebate or reduction.

The operative word deals with the "giving" of employment, and thus is directed to employers rather than employees. Accordingly, an engineering employer is barred not only from taking kickbacks from employees, but also from taking a payment in any form from an employment agency for "giving" employment to an engineer who secures a position through an employment agency. We assume that the reason for the restriciton on employment agencies is to prevent a secondary form of kickback arrangement in which the stated salary is reduced, in effect, by the employment agency charging the employee a substantial fee which reduces the actual income received by the employee.

That is not to say that an engineer seeking employment through an employment agency may not properly agree with the agency to pay it a commission for securing the position for the engineer. The bar applies to the employer taking any part of the commission paid by the employee to the agency as a condition of offering the employment.

The result of this reading of the Code provision is that it does not apply to Engineer A in the facts of this case. We have no doubt that in offering Engineer A an "acceptance bonus" for taking the position with Company Z, the employment agency is motivated by a desire to earn its fee from the employer.

If it thereby wishes to reduce the amount of its fee by the $2,000, it may do so as a business decision and is not controlled by the Code of Ethics. We perceive no ethical reason to prevent Engineer A from making a judgment on the type of employlment preferred, and from accepting an arrangement under these circumstances to enhance his economic interest through the payment which supports that interest consistent with the preferred employment.

Conclusion:
It is ethical for an engineer to accept a bonus payment from an employment agency as an inducement to accept employment with a particular employer.

10.18 Case Study 10.18

A) We noted in Case 69-2 that Section 11(d) of the code does not rule out all contingent contracts. Rather, Section 11(d) recognizes that contingent contracts are improper only under circumstances in which the arrangement may compromise the professional judgment of the engineer or when used as a device for promoting or securing a professional commission. The latter restriction is admittedly not as clear in its meaning as the first restriction. But we observed in Case 69-2 that we construed its purpose to be to safeguard the public and clients from projects which are unsound from a technical or economic standpoint. An example of the kind of restriction contemplated by the second condition is found in Case 65-4, in which we concluded that it would be unethical for an engineer to enter into a contingent contract under which his payment depends upon a favorable feasibility study for a public works project. We commented in that case: "The import of the restriction . . . is that the engineer must render completely impartial and independent judgment on engineering matters without regard to the consequences of his future retention or interest in the project." Following that premise in this case, there appears to be no possible or potential conflict between the interests of Engineer A and his clients through his search for lower utility rates on their behalf.

It is conceivable that Engineer A could compromise his professional judgment under these facts by overzealousness in seeking means of savings to his clients. But this kind of motivation can hardly produce a compromise of professional judgment, providing all other interests of the client, such as safety and reliability, are protected.

Whether this type of compensation arrangement is the most productive or wisest for Engineer A or his clients is not for us to say; it is enough to say that it is an arrangement not prohibited by the code.

Conclusion:

It is ethical for Engineer A to be compensated solely on the basis of a percentage of savings to his clients.

B) The Board has addressed the issue of contingent fees on numerous occasions. Recently in Case 81-1, we found that it would not be ethical for a firm to submit a contingent-contract proposal that included opinions as to the feasibility of the project. In that case, the proposal was accepted by the local government based upon the engineer's condition that the engineer would be given a letter of intent for the work stating that if the government secures the financing and proceeds with the project, a contract would be negotiated with the engineer, but otherwise the engineer would not be entitled to any fee or other payment. There the Board recognized that the engineer had been placed in a position of commitment and could not any longer be impartial with regard to the future analysis of the client's interest in proceeding or not proceeding with the project.

Similarly, in Case 65-14, the Board recognized the critical problem where an engineer's judgment might be influenced during the course of preliminary studies to produce a favorable finding that will result in the engineer's being retained for the full project. Said the Board, "The guiding principle in these kinds of cases is that the engineer must be careful not to include such degree of engineering services or opinions or conclusions on the economic and technical feasibility of the project that the engineer would run afoul of the restrictions imposed by the Code of Ethics."

Clearly, as the Baord noted in Case 73-4, Section III.7.a. "does not rule out all contingent contracts but rather recognizes that contingent contracts are improper . . . under circumstances in which the arrangement may compromise the professional judgment of the engineer or when used as a device for promoting a professional commission." We are of the view that the facts presented in the instant case are significantly different in two important respects from those cases which found that contingent contracts were improper on the ground that the engineer's judgment may be compromised. The facts here are materially different from those that were presented in Case 81-1. In that case the engineer aggressively volunteered a great deal of information to the local government. As noted in that case, the engineer went beyond the presentation of qualifications and sought to influence the client by volunteering certain information to show interest and desire for the project assignment and how the engineer would see the project's development. By contrast, in this case, Engineer A simply responded to the requests made by the city and did not go beyond the city's requests by engaging in promotional activities. Thus it would be much harder to demonstrate in this case that Engineer A's professional judgment might be compromised.

The second and more important distinction between this case and earlier cases involves the particular contract contingency and whether it may cause Engineer A's judgment to be compromised. We are of the view that the contingency involved in the instant case is not one in which

Engineer A's judgment may have cause to be compromised. Unlike those earlier cases in which particular contingencies related to whether a report demonstrated that a particular project was or was not feasible (see Cases 81-1 and 77-4), here the only contingency was whether a state agency responsible for funding the project would approve money from a general fund to the city.

The Board finds that there is nothing present under the facts of this case to indicate whether or not Engineer A would receive any type of preference or consideration in the selection of engineers for future design work in the event that the state angecy approved the funding and the city proceeded with the project. This Board therefore presumes that Engineer A had nothing to gain or to lose with respect to the contents of the report. Rather, this arrangement appears to be a purely speculative contract which the engineering firm agreed to without receiving any future considerations. Thus, Engineer A agreed to perform certain services for the client on a speculative basis. As this Board noted in Case 77-4, "...we do not find that the Code by its specific language bars an engineer from entering into a purely speculative contract. If the engineer wishes to take the chance, ...he or she may do so."

Conclusion:
It was not unethical for Engineer A to perform the requested services under the contingency contract between Engineer A and the local government.

10.19 Case Study 10.19

The Board of Ethical Review found that it was ethical for Engineers A and B to offer conflicting opinions on the application of engineering principles and for each engineer to criticize the work of the other at hearings. The criticism, however, must be in the public interest and must be given with a high level of professional deportment.

Some aspects of an engineering problem will admit of only one conclusion, such as the solution to a mathematical equation, but that doesn't mean all engineering problems permit only one correct answer. Large public projects are often in that "inexact" category. The approach to such projects may properly reflect not only engineering diagnoses but also questions of public policy. Honest differences of opinion among equally qualified engineers on the interpretation of the known physical facts may also exist.

In publicly criticizing the work of another engineer, however, engineers should avoid any form of personal attack and should base their criticisms strictly on the engineering conclusions or application of data, while offering alternative conclusions or analysis.

10.20 Case Study 10.20

The Board of Ethical Review operates on an "ad hoc" educational basis, and does not engage in resolving disputes of fact between parties in actual cases. That function is left to the state society if members are involved in judging whether a member has violated the Code of Ethics. Being solely educational, the function of the Board of Ethical Review is to take the submission of "facts" as the basis for analysis and opinion without attempting to obtain rebuttal or comment from other parties. On that basis, the reader of the opinions should always recognize that the Board of Ethical Review is not an adjudicatory body, but its opinions are intended to apply to actual cases only to the extent of the "facts", stated in the case.

This case presents a series of facts, some of which may be addressed by the Board of Ethical Review, others that may not. It appears from the facts that certain wrongdoings were committed by a non-engineer. However, the Board of Ethical Review does not review the conduct of non-engineers with respect to the Code of Ethics. Non-engineers, of course, are not covered by the

Code and therefore it would be a meaningless act for this Board to review the conduct of Smith in the facts presented above. Instead, it is the duty of the Board to focus upon the actions of Engineer A.

In Case 64-7, the Board interpreted Section III.10.a. (then Sections 14 and 14(a)) to mean that individual accomplishments and the assumption of responsibility by individual engineers should be recognized by other engineers. "This principle," said the Board, "is not only fair and in the best interests of the profession, but it also recognizes that the professional engineer must assume personal responsibility for decisions and actions." Although the facts of that case were somewhat different from those in the case at hand, Case 64-7 reflects the view that each individual engineer has an ethical obligation to recognize and give credit to the creative products of other engineers. At a bare minimum, that ethical obligation includes securing the consent of that engineer, indicating on any reproduction of that creation the identity of the engineer and in some cases providing the engineer with remuneration for his work depending upon the surrounding circumstances. Each case must be decided upon its individual facts, as no two cases are alike. However, certain basic obligations exist that must be recognized in all cases.

If in fact Engineer A used the proposal, it is clear that such a use would be in violation of Section III.10.a. of the Code of Ethics. Although it may be argued that the Code provision is meant to address those situations where a supervising engineer fails to give credit to an employee responsible for a particular design, and not where "proposals" (which might in fact even be a matter of public record) are submitted by several firms and one engineer merely reviews another set of proposals to gain another firm's perspective of the project, we are convinced that the Code may properly be read to imply use and thus proscribe the conduct of Engineer A. The Board concludes from the facts that the general purpose of Engineer A's use of the proposal of Engineer B was to develop a proposal and thus be awarded the contract. That being the purpose, Engineer A had an obligation to (1) seek and obtain Engineer B's consent before using the plans as a basis for one's own proposal; (2) if granted consent, identify Engineer B in all cases of use of Engineer B's proposal; and (3) negotiate and pay Engineer B "fair and reasonable" compensation for using the proposal. By failing to fulfill any of those obligations, Engineer A clearly violated Sections III.10. and III.10.a. of the Code.

The actions of Engineer A suggest conduct unbecoming of a professional engineer. When offered the contents of Engineer B's proposal by Smith, Engineer A had an ethical obligation to refuse to accept the proposal. Instead, Engineer A accepted and also used the material. Because of the decision to actually use the material, we must further conclude that Engineer A violated Section III.7. of the Code by competing unfairly with Engineer B by attempting to "obtain...advancement...by...improper or questionable methods." Although that Code provision is broad and leaves a good deal of room for interpretation, we are convinced that the use of the proposal constituted unfair competition by improper and questionable methods. Whether there would have been a violation of Section III.7. had Engineer A not used Engineer B's proposal but merely reviewed it before developing the proposal is a debatable point that we will leave for another day. However, this Board is being asked to determine whether a violation occurred as a result of Engineer A's use of Engineer B's proposal. We think that Engineer A's use under the present facts constitutes unfair competition by improper and questionable methods and hence a violation of Section III.7. of the Code.

Conclusion:
It was unethical for Engineer A to use Engineer B's proposal without Engineer B's consent in order to develop a proposal that was subsequently submitted to the council.

10.21 Case Study 10.21

On prior occasions, this Board has reviewed the issue of "honesty in academic endeavors." While the facts of those situations are quite a bit different from the facts in the instant case, and probably somewhat more clear-cut, we believe it is useful to review the cases in order to gain a full appreciation of the issues present in this case.

In BER Case 75-11, the Board reviewed a situation involving an engineer, Engineer #1, who performed certain research and then prepared a paper on an engineering subject based on that research which was duly published in an engineering magazine under his byline. Subsequently, an article on the same subject written by Engineer #2 appeared in another engineering magazine. A substantial portion of the text of Engineer #2's article was identical, word for word, with the article authored by Engineer #1. Engineer #1 contacted Engineer #2 and requested an explanation. Engineer #2 replied that he had submitted with his article a list of six references, one of which identified the article by Engineer #1, but that the list of references had been inadvertently omitted by the editor. Engineer #2 offered his apology to Engineer #1 for the mishap because his reference credit was not published as intended.

Not the least bit surprisingly, the Board ruled Engineer #2's conduct not in accord with the Code of Ethics. The Board, noting that "this is a clear case of plagiarism and [is] directly offensive to the Code," indicated that "merely listing the work of Engineer #1 in a list of references to various articles only tells the reader that Engineer #2 had consulted and read those cited articles of other authors. It in no way tells the reader that a large portion of his text is copied from the work of another."

While we in no way suggest that the facts of BER Case 75-11 are analogous to those of the instant case, we do believe they suggest the vital importance of "honesty in academic endeavors," and the confusion and distortion that arise when one fails to strive toward that end.

A second case relating to the issue of academic honesty relates to the subject of academic qualifications. In BER Case 79-5 Engineer A received a B.S degree in 1940 from a recognized engineering curriculum and subsequently was registered as a professional engineer in two states. Later, he was awarded an earned "Professional Degree" from the same institution. In 1960 he received a Ph.D. degree from an organization that awarded degrees on the basis of correspondence without requiring any formal attendance or study at the institution, and was regarded by state authorities as a "diploma mill." Engineer A listed his Ph.D. degree among his academic qualifications in brochures, correspondence, etc., without indicating its nature.

The Board found that Engineer A was unethical in citing his Ph.D. degree as an academic qualification under those circumstances, noting that "Engineer A is charged with knowledge of the accepted standards of the profession. In stating that he had a Ph.D. degree, he should have been aware that those who received his communications would be deceived."

Those two cases, although quite a bit different from the case at hand, are extremely useful in understanding the vital importance of honesty in academic endeavors, and particularly in the field of engineering research. While at first blush, those two cases do not appear to present particularly crucial issues involving honesty in academic endeavors, they do suggest an important point. Both cases reveal what could probably best be described as a kind of "intellectual laziness" on the part of the engineers in question. Both are fairly simple cases: An engineer who engages in plagiarism is not ethical. Nor is an engineer who tries to puff up his credentials with a degree secured through a "diploma mill" ethical.

But what about the instant case? Is an engineer who fails to include unsubstantiative data in his graduate report unethical? In view of the fact that no BER decisions have heretofore examined this question, it is necessary for the Board to examine the pertinent portions of the Code of Ethics.

We think that Section II.3.a. is a good starting point. That provision unambiguously enunciates the ethical duty of the engineer in this area. The engineer must be objective and truthful in his professional reports and must include all relevant and pertinent information in such reports. In the instant case, that would suggest that Engineer A had an ethical duty to include the unsubstantiative data in his report because such data were relevant and pertinent to the subject of his report. His failure to include them indicates that Engineer A may have exercised subjective judgment in order to reinforce the thrust of his report.

Section III.3.a. is also relevant to our inquiry. In a sense, Engineer A's failure to include the unsubstantiative data in his report caused his report to be somewhat misleading. An individual performing research at some future date, who relies upon the contents of Engineer A's report, may assume that his results are unqualified, uncontradicted, and fully supportable. That may cause such future research to be equally tainted and may cause future researchers to reach erroneous conclusions.

Finally, we believe that Section III.11. should play a part in our discussion. We do not see how Engineer A could be acting consistently with that provision by failing to include the unsubstantiative data in his report. By misrepresenting his findings, Engineer A distorts a field of knowledge upon which others are bound to rely and also undermines the exercise of engineering research. Although Engineer A may have been convinced of the soundness of his report based upon his overall finding and concerned that inclusion of the data would detract from the thrust of his report, such was not enough of a justification to omit reference to the unsubstantiative data. The challenge of academic research is not to develop accurate, consistent, or precise findings which one can identify and categorize neatly, nor is it to identify results that are in accord with one's basic premise. The real challenge of such research is to wrestle head-on with the difficult and sometimes irresolvable issues that surface, and try to gain some understanding of why they are at variance with other results.

Conclusion:
It was unethical for Engineer A to fail to include reference to the unsubstantiative data in his report.

10.22 Code of Ethics for Engineers

Engineering is an important and learned profession. The members of the profession recognize that their work has a direct and vital impact on the quality of life for all people. Accordingly, the services provided by engineers require honesty, impartiality, fairness and equity, and must be dedicated to the protection of the public health, safety and welfare. In the practice of their profession, engineers must perform under a standard of professional behavior which requires adherence to the highest principles of ethical conduct on behalf of the public, clients, employers and the profession.

- I. Engineers, in the fulfillment of their professional duties, shall:

  1. Hold paramount the safety, health and welfare of the public in the performance of their professional duties.
  2. Perform services only in areas of their competence.
  3. Issue public statements only in an objective and truthful manner.
  4. Act in professional matters for each employer or client as faithful agents or trustees.
  5. Avoid improper solicitation of professional employment.

• II. RULES OF PRACTICE

1. Engineers shall hold paramount the safety, health and welfare of the public in the performance of their professional duties.

    a. Engineers shall at all times recognize that their primary obligation is to protect the safety, health, property and welfare of the public. If their professional judgment is overruled under circumstances where the safety, health, property or welfare of the public are endangered, they shall notify their employer or client and such other authority as may be appropriate.

    b. Engineers shall approve only those engineering documents which are safe for public health, property and welfare in conformity with accepted standards.

    c. Engineers shall not reveal facts, data or information obtained in a professional capacity without the prior consent of the client or employer except as authorized or required by law or this Code.

    d. Engineers shall not permit the use of their name or firm name nor associate in business ventures with any person or firm which they have reason to believe is engaging in fraudulent or dishonest business or professional practices.

    e. Engineers having knowledge of any alleged violation of this Code shall cooperate with the proper authorities in furnishing such information or assistance as may be required.

2. Engineers shall perform services only in the areas of their competence.

    a. Engineers shall undertake assignments only when qualified by education or experience in the specific technical fields involved.

    b. Engineers shall not affix their signatures to any plans or documents dealing with subject matter in which they lack competence, nor to any plan or document not prepared under their direction and control.

    c. Engineers may accept an assignment outside of their fields of competence to the extent that their services are restricted to those phases of the project in which they are qualified, and to the extent that they are satisfied that all other phases of such project will be performed by registered or otherwise qualified associates, consultants, or employees, in which case they may then sign the documents for the total project.

3. Engineers shall issue public statements only in an objective and truthfull manner.

    a. Engineers shall be objective and truthful in professional reports, statements or testimony. They shall include all relevant and pertinent information in such reports, statements or testimony.

    b. Engineers may express publicly a professional opinion on technical subjects only when that opinion is founded upon adequate knowledge of the facts and competence in the subject matter.

    c. Engineers shall issue no statements, criticisms or arguments on technical matter which are inspired or paid for by interested parties, unless they have prefaced their comments by explicitly identifying the interested parties on whose behalf they are speaking, and by revealing the existence of any interest the engineers may have in the matters.

4. Engineers shall act in professional matters for each employer or client as faithful agents or trustees.

    a. Engineers shall disclose all known or potential conflicts of interest to their employers or clients by promptly informing them of any business association, interest, or other circumstances which could influence or appear to influence their judgment or the quality of their services.

b. Engineers shall not accept compensation, financial or otherwise, from more than one party for services on the same project, or for services pertaining to the same project, unless the circumstances are fully disclosed to, and agreed to, by all interested parties.

c. Engineers shall not solicit or accept financial or other valuable consideration, directly or indirectly, from contractors, their agents, or other parties in connection with work for employers or clients for which they are responsible.

d. Engineers in public service as members, advisors, or employees of a governmental body or department shall not participate in decisions whith respect to professional services solicited or provided by them or their organizations in private or public engineering practice.

e. Engineers shall not solicit or accept a professional contract from a governmental body on which a principal or officer of their organization serves as a member.

5. Engineers shall avoid improper solicitation of professional employment.

a. Engineers shall not falsify or permit misrepresentation of their, or their associates', academic or professional qualifications. They shall not misrepresent or exaggerate their degree of responsibility in or for the subject matter of prior assignments. Brochures or other presentations incident to the solicitation of employment shall not misrepresent pertinent facts concerning employers, employees, associates, joint venturers or past accomplishments with the intent and purpose of enhancing their qualifications and their work.

b. Engineers shall not offer, give, solicit or receive, either directly or indirectly, any political contribution in an amount intended to influence the award of a contract by public authority, or which may be reasonably construed by the public of having the effect or intent to influence the award of a contract. They shall not offer any gift, or other valuable consideration in order to secure work. They shall not pay a commission, percentage or brokerage fee in order to secure work except to a bona fide employee or bona fide established commercial or marketing agencies retained by them.

- III. PROFESSIONAL OBLIGATIONS

1. Engineers shall be guided in all their professional relations by the highest standards of integrity.

a. Engineers shall admit and accept their own errors when proven wrong and refrain from distorting or altering the facts in an attempt to justify their decisions.

b. Engineers shall advise their clients or employers when they believe a project will not be successful.

c. Engineers shall not accept outside employment to the detriment of their regular work or interest. Before accepting any outside employment they will notify their employers.

d. Engineers shall not attempt to attract an engineer from another employer by false or misleading pretenses.

e. Engineers shall not actively participate in strikes, picket lines, or other collective coercive action.

f. Engineers shall avoid any act tending to promote their own interest at the expense of the dignity and integrity of the profession.

2. Engineers shall at all times strive to serve the public interest.

a. Engineers shall seek opportunities to be of constructive service in civic affairs and work for the advancement of the safety, health and well-being of their community.

b. Engineers shall not complete, sign, or seal plans and/or specifications that are not of a design safe to the public health and welfare and in conformity with accepted engineering standards. If the client or employer insists on such unprofessional conduct, they shall notify the proper authorities and withdraw from further service on the project.

c. Engineers shall endeavor to extend public knowledge and appreciation of engineering and its achievement and to protect the engineering profession from misrepresentation and misunderstanding.

3. Engineers shall avoid all conduct or practice which is likely to discredit the profession or deceive the public.

   a. Engineers shall avoid the use of statements containing a material misrepresentation of fact or omitting a material fact necessary to keep statements from being misleading; statements intended or likely to create an unjustified expectation; statements containing prediction of future success; statements containing an opinion as to the quality of the Engineers' services; or statements intended or likely to attract clients by the use of showmanship, puffery, or self-laudation, including the use of slogans, jingles, or sensational language or format.

   b. Consistent with the foregoing, Engineers may advertise for recruitment of personnel.

   c. Consistent with the foregoing, Engineers may prepare articles for the lay or technical press, but such articles shall not imply credit to the author for work performed by others.

4. Engineers shall not disclose confidential information concerning the business affairs or technical processes of any present or former client or employer without his consent.

   a. Engineers in the employ of others shall not, without the consent of all interested parties, enter promotional efforts or negotiations for work or make arrangements for other employment as a principal or to practice in connection with a specific project for which the Engineer has gained particular and specialized knowledge.

   b. Engineers shall not, without the consent of all interested parties, participate in or represent an adversary interest in connection with a specific project or proceeding in which the Engineer has gained particular specialized knowledge on behalf of a former client or employer.

5. Engineers shall not be influenced in their professional duties by conflicting interests.

   a. Engineers shall not accept financial or other considerations, including free engineering designs, from material or equipment supplies for specifying their product.

   b. Engineers shall not accept commissions or allowances, directly or indirectly, from contractors or other parties dealing with clients or employers of the Engineer in connecton with work for which the Engineer is responsible.

6. Engineers shall uphold the principle of appropriate and adequate compensation for those engaged in engineering work.

   a. Engineers shall not accept remuneration from either an employee or employment agency for giving employment.

   b. Engineers, when employing other engineers, shall offer a salary according to professional qualifications and the recognized standards in the particular geographical area.

7. Engineers shall not compete unfairly with other engineers by attempting to obtain employment or advancement of professional engagements by taking advantage of a salaried position, by criticizing other engineers, or by other improper or questionable methods.

   a. Engineers shall not request, propose, or accept a professional commission on a contingent basis under circumstances in which their professional judgment may be compromised.

    b. Engineers in salaried positions shall accept part-time engineering work only at salaries not less than that recognized as standard in the area.

    c. Engineers shall not use equipment, supplies, laboratory, or office facilities of an employer to carry on outside private practice without consent.

8. Engineers shall not attempt to injure, maliciously or falsely, directly or indirectly, the professional reputation, prospects, practice or employment of other engineers, nor indiscriminately criticize other engineers' work. Engineers who believe others are guilty of unethical or illegal practice shall present such information to the proper authority for action.

    a. Engineers in private practice shall not review the work of another engineer for the same client, except with the knowledge of such engineer, or unless the connection of such engineer with the work has been terminated.

    b. Engineers in governmental, industrial or educational employ are entitled to review and evaluate the work of other engineers when so required by their employment duties.

    c. Engineers in sales or industrial employ are entitled to make engineering comparisons of represented products with products of other suppliers.

9. Engineers shall accept personal responsibility for all professional activities.

    a. Engineers shall conform with state registration laws in the practice of engineering.

    b. Engineers shall not use association with a nonengineer, a corporation, or partnership, as a "cloak" for unethical acts, but must accept personal responsibility for all professional acts.

10. Engineers shall give credit for engineering work to those to whom credit is due, and will recognize the proprietary interests of others.

    a. Engineers shall, whenever possible, name the person or persons who may be individually responsible for designs, inventions, writings, or other accomplishments.

    b. Engineers using designs supplied by a client recognize that the designs remain the property of the client and may not be duplicated by the Engineer for others without express permission.

    c. Engineers, before undertaking work for others in connection with which the Engineer may make improvements, plans, designs, inventions, or other records which may justify copyrights or patents, should enter into a positive agreement regarding ownership.

    d. Engineers' designs, data, records, and notes referring exclusively to an employer's work are the employer's property.

11. Engineers shall cooperate in extending the effectiveness of the profession by interchanging information and experience with other engineers and students, and will endeavor to provide opportunity for the professional development and advancement of engineers under their supervision.

    a. Engineers shall encourage engineering employees' efforts to improve their education.

    b. Engineers shall encourage engineering employees to attend and present papers at professional and technical society meetings.

    c. Engineers shall urge engineering employees to become registered at the earliest possible date.

    d. Engineers shall assign a professional engineer duties of a nature to utilize full training and experience, insofar as possible, and delegate lesser functions to subprofessionals or to technicians.

e. Engineers shall provide a prospective engineering employee with complete information on working conditions and proposed status of employment, and after employment will keep employees informed of any changes.

Appendix B

Hints for Design Projects

B.1 Design Project B.1

The possibility of using Freon 11 as the working fluid in a gravity actuated heat engine was evaluated. The principal of operation used was similar to that used by the "dunking bird" toy. A model was developed and tested by this group. The power developed by the system was measured to be 0.25 Watts, which was considered to be insufficient for agricultural irrigation needs. At an angular velocity of 0.167 rpm the maximum torque measured was 10.6 lbf-ft. The system was unstable in operation. When the wheel was allowed to rotate freely the full (heavy) pods rotated into the bath and began immediately bleeding fluid into the upper pods. The center of mass moved to a point well above the axis, and the system rotated swiftly (10 to 20 rpm) to a more stable position with the center of mass located below the axis. With some sort of damping it is thought that the engine could be stabilized.

B.2 Design Project B.2

This project involves a high pressure gaseous nitrogen (GN2) pressure vessel that was used in the Thor missile system during the late 1950's and early 1960's. The vessel used a multiwall design concept with fabrication by the A. O. Smith Corporation, a company more notable for hot water heater fabrication than for high pressure vessels. It would probably be helpful for students assigned to this project to have previously worked Problem A.1, which requires the development of a drawing system for this vessel and the associated components. Since the components shown attached to the vessel in Figure B.1(a) and (b), are not shipped installed, the installation procedure must provide instructions as to how this installation is accomplished in the field. The cleanliness of the system is critical and therefore the installation procedure must also insure that system cleanliness is maintained. This is usually accomplished by keeping the system closed (protected from the environment) to the maximum extent possible and by providing a low pressure purge gas of GN2 at the pressure vessel and allowing the gas to flow out of the system when connections are made thus keeping the environment from entering the system. Plastic enclosures can also be constructed over the work area to further resist intrusion by the environment.

The installation drawing should show the vessel mounted on a concrete pad at the proper elevation and properly secured using bolts or studs. All components shipped with the vessel should be installed by the installation procedure. A blind flange should be installed at the termination and the pressure vessel and system should be maintained under a low blanket or purge pressure of GN2 (approximately 10 psi).

B.3 Design Project B.3

The key to this project is recognizing that the FOVMA must be installed in parallel to the main fuel line (not directly in it), with periodic or continuous sampling of the oil viscosity. Several techniques have been suggested for measuring the viscosity of fuels and several methods have proved to be very successful in the laboratory. However, these laboratory techniques are usually not suitable for field use where a rugged and reliable device is required. Some methods measure the velocity in a known diameter capilliary tube whereas others have measured the speed of counterrotating devices where the fuel is located in the annular space between the two devices. Almost any solution will have to incorporate a small computer and the necessary control system to control the heater in the main fuel line.

B.4 Design Project B.2

In this project several approaches are viable as to the source of the natural gas fuel. Many metropolitan areas now have public refueling facilities that sell natural gas for approximately $0.70/gallon (equivalent), which may be the least costly approach for obtaining a supply of fuel. Refueling facilities are expensive to construct, primarily because of the compressor, which can vary in cost from approximately $3,000 to $100,000, depending on its capacity. If the facility can be constructed reasonably close to a medium or high pressure natural gas line the cost of the compressor can be reduced. Another large cost item in refueling facilities is the degree of automated data collection required. If card operated dispensing systems are required, the cost of the facility may be increased by $50,000 to $100,000. More often than not, this level of sophistication is not required, at least not initially. Another cost driver is the decision as to whether fastfill capability is required. For many fleet operations fastfill is not required and thus the cost of the pressure vessel cascade can be eliminated.

Another area of consideration is the type of conversion that will be used on the vehicles. Electronic systems are now becoming available with better control and consequently better emissions and possibly better fuel economy. Unfortunately, they cost more than the mechanical systems and thus will take longer to pay off through fuel savings.

B.5 Design Project B.5

This is a good machine component design project in that it integrates much of the machine component design theory discussed in early mechanical engineering design courses. The problems that students encounter in the speed reducer design project are primarily associated with tolerances, interference fits and ability to assemble the housing with all the components. Students often do not provide a means for the housing to be assembled that will allow maintaining the interference fits required for the bearings. Often the selection of a proper lubricating oil (which should be heavy weight), and means to add and drain the oil from the housing, is omitted. Installation of the oil seal (which should be press fitted into the housing) is also often not provided for. In designing the gear the student should make calculations to verify that no interference exists and no undercutting is required.

B.6 Design Project B.6

- Calculate the required pump hp.

- Determine the engine speed.

- Calculate the volume swept out by one stroke.

- Calculate the dimensions of piston and piston stroke for the design.

- Assume the *clearance volume* to be about %5 of the total volume swept out.

- Select cast iron as the material for the piston and cylinder [Potter, Andrey A., "Elements of Steam and Gas Power Engineering," MacGraw-Hill Book Comp., Inc., London, 1924].

- Calculate the cylinder thickness based on ASTM 60 cast iron which has an ultimate tensil strength of 60 kpsi.

- The area where the piston rod leaves the crankend cylinder head is made steam tight by using a stuffing box. Babbitt metal rings pressed up against the piston rod would be a good suggestion for the metallic packing in the stuffing bax [Shealy, E. M., "Steam Engines," MacGraw-Hill Book Comp., Inc., London, 1919].

- The joint between the cylinder head and the cylinder is made steam tight with the use of a gasket seal. A fluorelastomer would be a good material to make the gasket seal [Dahlheimer, John C., "Mechanical Face Seal Handbook," Chilton Book Comp., Philadelphia, 1982].

- Select a simple slide valve with laps [Shealy, E. M., "Steam Engines," MacGraw-Hill Book Comp., Inc., London, 1919]. The slide valve should produce a cut-off of 1/3 of the stroke for this particular design.

B.7 Design Project B.7

Any closed circuit testing system or "tunnel" has four major components regardless of the fluid being employed. These components are:

1. A contoured duct to control and direct fluid flow.

2. A drive system to move the fluid through the duct.

3. A model of the object to be tested.

4. Instrumentation to measure forces, moments, and pressure exerted on the model by the moving fluid.

The test section is the most important portion of the tunnel and it governed the design of the entire system. The test section as designed was to have water flowing at 1 foot per second through a 1 square foot cross sectional area. This requires a volumetric flow rate of 1 cubic foot per second, or 450 gallons per minute. In order to view the model being tested, the test section walls were to be made of glass or a polymeric substitute. The test section was designed to have a length of three feet in order to allow viewing of vortex shedding around the model. Glass was selected for its extremely low friction factor, thus minimizing boundary layer effects in the section. A glass thickness of 3/8 inch was selected to provide a minimum safety factor of 4.0 under the nominal static and dynamic pressure of just over 1 foot of water or 0.5 pounds per square inch. As a matter of general practice, the maximum model size which could be tested in a tunnel of this design would represent a total flow blockage of 10 percent. Thus, the frontal area of any model placed in this system for testing could be no larger than 14.5 square inches. The reason for this 10 percent limit is to prevent excessive flow acceleration around the model from distorting the results. The top of the test section was specified to be removable in order to facilitate model removal, insertion, and adjustment without requiring the draining of the system. It should also be possible to run the system either at low speeds or with the system 90 percent full while the top is removed.

Dead section and exit plenum

As the flow exited the test section, it was designed to pass through a 2 foot long and 1 foot square cross section zone, which served to isolate the test section from the exit plenum. To a certain degree, back flow and swirl are expected from the 90 degree bend of the exit plenum and the entrance into the return loop piping system. Beyond the dead section, the exit plenum was designed to turn the flow downward by 90 degrees.

On the exit side of the exit plenum, four straight vanes were specified to prevent the swirl effects from the pipe entrance from working back toward the test section. The pipe entrance was designed to be a smooth transition from a 1 foot by 1 foot cross section to a 3 inch pipe diameter rather than a sharp pipe entrance.

Main and Transition Diffusers

The first element of the tunnel after the piping system is a transition diffuser, which was designed to be a mirror image of the smooth transition at the entrance to the plumbing system. This diffuser was designed to provide some dispersion of the flow stream that would not be possible with a sharp entrance through a one foot square plate.

The design of the main diffuser is governed by the test section size, test section aspect ratio, and the tunnel contraction ratio. The diffuser was required to expand the flow from a 1 foot square cross-section leaving the transition diffuser to five times the test section area, or five square feet. A planar rectangular diffuser was selected for this design.

The diffuser was to provide a transition from one cross-sectional area to another, but also to provide a fairly even velocity profile at the exit of the section. This second requrement is the more difficult one to implement.

In order to provide an even velocity profile, the diffuser must avoid the condition known as stalling. Stalling occurs in a diffuser when the velocity and diffusion angle exceed experimentally determined limits beyond which, the boundary layer along the diverging wall will separate and the flow will back flow and eddy. When stalling occurs. severe vibrations can be produced in the diffuser. In order to avoid stalling in a rectangular diffuser, the diffusion angle must be less than 12 degrees.

Straightening section

A 3 feet long, 5 feet wide, and 1 foot high rectangular section was added to the system after the main diffuser. While the diffuser would provide an even velocity profile, it would be less than perfect. The straightening section contains screens, grids, honeycombs, or perforated plates, to even out small velocity profile differences and induce a uniform turbulence field. The number and types of devices must be determined by experimentation. The nominal velocity in this section was designed to be 0.2 feet per second.

Entrance cone

Once the velocity and turbulence fields have been induced, the flow must then be contracted from the 5 square feet cross section to the 1 square foot cross section of the test section. The entrance cone must accomplish this while maintaining uniformity of the velocity and turbulence fields. The entrance cone was selected with dimensions similar to the diffuser. In order to maintain the flow fields, the walls of the entrance cone are curvilinear. This shape is standard for aerodynamic tunnel design. Like the diffuser and straightening section, the entrance cone only acts in a horizontal plane and has a uniform depth of 1 foot.

Piping and pumping system

From the specified flow rate through the system and an assumption as to the number and types of straightening devices that would be required, the maximum dynamic pressure in the tunnel would be at the junction of the transition diffuser and main diffuser. Less than 1/2 foot of water, or 0.2 pounds per square inch of dynamic head is expected at this point.

This design called for either an axial flow pump or a high flow rate, low head centrifugal pump. At the design flow rate, a pump head of 15 feet is expected, and in order to reduce this to 1/2 foot of water head required at the tunnel entrance, a pipe diameter of 3 inches was specified for the 30 feet of return piping with a valve to provide further flow rate and pressure control.

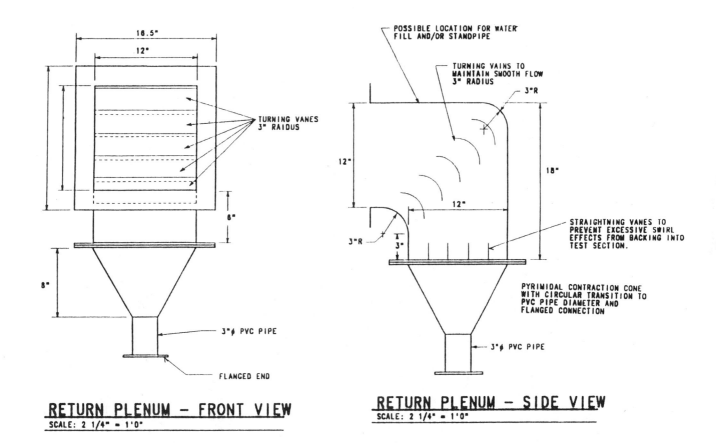

Side and top view of water tunnel

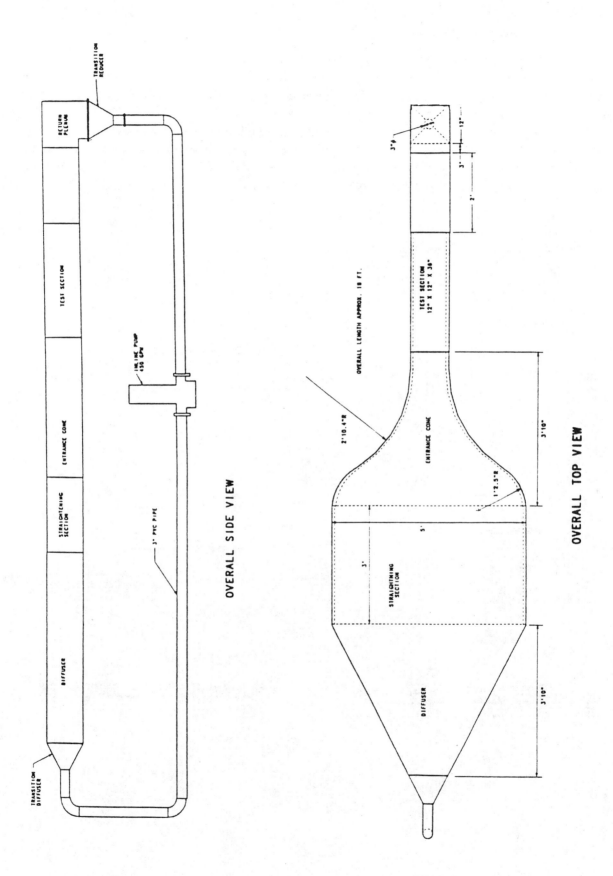

110

B.8 Design Project B.8

Follow the procedure outlined in Design Project B.7.

B.9 Design Project B.9

There are two principle construction standards in general use, the American Petroleum Institute (API) Standard 650 covering "Welded Steel Tanks For Oil Storage," and the American Water Works Association (AWWA) covering "Steel Tanks for Water Storage." Both API and AWWA approve the use of a wide variety of steel plate material. The AWWA-D100 is a public standard to be used without reservation for the storage of a single product, water. AWWA designs are based on the stress existing at the lower edge of each course. The AWWA-D100 standard was used for the design analysis of this project.

The analysis of this project showed the family-of-steels design method to be a more efficient and less expensive solution to the problem. By incorporating a series of 4 steels as shown in Figures B.9.1 and B.9.2, including ASTM A283-C, A441, A656, A517-F("T-1"), the family-of-steels method resulted in a weight reduction of 142,528 lb and an initial material cost reduction of $85,172. It should be noted that in addition to the savings in material costs, money will also be saved in shipping and handling of material due to the significant reduction in total weight. Moreover, the fabrication, erection, and welding costs should also be reduced by the family-of-steels design.

Although the family-of-steels method reduced the costs of the tank, there are other aspects to consider. For instance, some of the materials selected for this design may be hard to obtain. Also, when joining various types of steel welding compatibility is a concern.

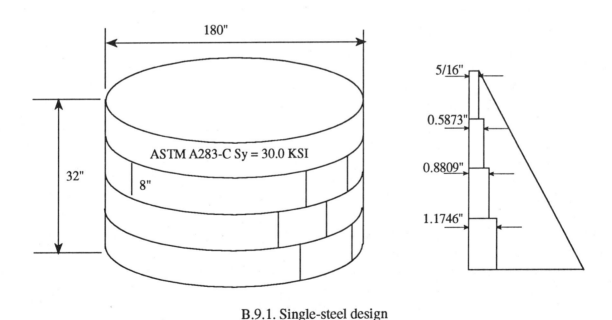

B.9.1. Single-steel design

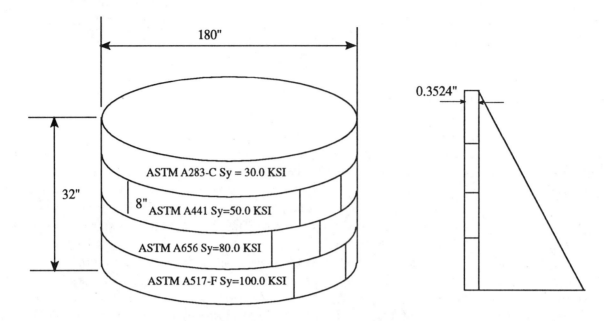

B.9.2. Family-of-steels design

B.10 Design Project B.10

Refer to the hints provided for design project B.9.

B.11 Design Project B.11

In accordance with the goals outlined in the project description an Automated Packaging for Plastic Eggs device was designed and developed.

Technical Description

To automate the egg sorting and packaging process, storage bins with motor-driven rotary disks were interfaced with a computer control system including photocells and associated circuitry. Nine 2x2x3 foot storage bins capable of storing 1500 eggs were constructed of sheet metal. The eggs are delivered to a rotary disk feeder at the bottom of the storage bin by use of an eccentrically located funnel that allows only one egg to be dropped from the bin at any one time. The 18-inch diameter disk is driven by an electric motor. Each disk has eight 2.5 inch slots which allow the egg to drop into the opening and be rotated until aligned with the delivery chute. The delivery chute is an open chute system constructed of sheet metal which converges to a central location where cartons are filled with the eggs. The entire structure is elevated 3.5 feet above the floor level to provide space for the gravity chute system to be installed. A platform is provided to allow filling the bins with eggs.

The electrical system consists of a computer, associated circuitry, photocells and electric motors. The computer is a 64K, single disk drive Apple II personal computer. Computer input and output is controlled by a program written in BASIC language. The computer provides signals through its game port to a circuit connected to a 5 and 12 volt power supply. The photocells are located at the bin outlet to indicate when an egg has dropped. The photocells are prefabricated light sensors that include associated circuitry that stops the motor as soon as the beam is broken. The motors are 110 volt, 6 rpm Dayton electric motors. Upon receipt of a 12 volt impulse by the photocell after computer system activation the motor driving the rotating disks is activated. When an egg drops, breaking the photocell beam, the motor is turned off, signalling the computer that one egg has been delivered. The computer then reactivates the motor through the photocell and repeats the entire sequence until the required number of eggs has been delivered. The circuitry includes 18 transistors, 2 multiplexers, and 2 demultiplexers connected to the 5 and 12 volt power supply. Total system cost was $2445 (only hardware).

System Operation

The capability of the system is approximately 30 boxes of eggs per hour which corresponds to 300,000 eggs per week, a significant increase over the 80,000 eggs per week achieved prior to installation of the system. Handicaped students are successfully operating the system using the computer interface and thus have relieved workshop supervision to concentrate on other tasks. Most students, regardless of their skill level, have been able to operate the system. Additional income for students has been generated from the plastic egg contract, as well.

B.12 Design Project B.12

Design Project B12

In response to this need a simple conduit frame was constructed upon which the paper rolls could be placed for processing. A moveable section of conduit that swiveled on the frame was included to install and remove the paper rolls. For paper rolls of larger diameter an adapter that slipped over the moveable section of conduit was provided. A simple cutter was constructed from a rain gutter drip section which had a convoluted edge that made cutting the paper relatively easy and yet, was not sharp enough to cut the operator's hands. The cutter was attached to the frame about 8 inches below the paper roll.

Testing of the device proved very successful and three more of the paper cutters were constructed for use at the school as a result. Cost of the cutter was $30 each.

B.13 Design Project B.13

In accordance with the goals outlined in the project description an automated can crushing device was developed. The design process is described below.

Design Criteria and Objectives

The design criteria were driven by the limited capabilities of the students for whom the device was developed. Some of the students had such limited mobility that they were only capable of moving their elbow. Thus, this lack of mobility dictated that the machine be capable of activation from a single, easily contacted button. The machine was also required to be capable of operation without the continuous aid of a teacher or supervisor. The can crusher needed to be safe, effective, and economical in performing the following actions:

1. Loading the can automatically.
2. Crushing the can on command.
3. Discharging the crushed can.
4. Stopping the cycle after the can was crushed.

For the machine to be safe it had to be constructed so that the students could not get a hand or finger in the mechanism. Finally, the crusher needed to be structurally sound, reliable and easy to maintain.

The Design Process

To satisfy the above criteria the design concept adopted required the crushing process to be initiated by an easily activated button switch. This action drove a can crushing mechanism which would crush the can to approximately 1- inch in height, eject the crushed can into a receiving container, and then return to its inactivated position. In moving back to the inactive position a triggering device released a second can from a holding bin. This can dropped into the crushing container in preparation for the next process to be initiated.

One of the first steps in the design process was to determine the type of press to use for the can crushing action. The viable possibilities included an electric power screw, a pneumatic cylinder or an hydraulic cylinder. Initially, it was thought that the power screw would be the best selection due to safety, noise and facility requirements considerations. However, after considerable analysis and discussion with suppliers this option was eliminated due to cost. The hydraulic cylinder was also eliminated due to cost, maintenance and the likelihood of spills, leaks, etc. The pneumatic cylinder was selected as the best choice for this application due to its low cost, ease of maintenence and ability to achieve relatively high forces at reasonably low pneumatic pressures. The pneumatic cylinder approach was lowest in cost even though a compressor had to be purchased to supply the air pressure.

Because the equations applicable to short compression members were extremely difficult to apply to soft drink cans, a compression test was performed to determine the force necessary to crush the can. This test was performed using a mechanical strain frame. The maximum force required to crush the can to an approximate height of 1 inch was determined to be 400 lbf.

Based on this information detail design was initiated. The can crusher consists of three subsystems, viz. the can crushing container, the can holding bin, and the control subsystem. In addition, an air compressor and storage tank capable of providing 80 to 90 psi pressure was provided.

The Can Crushing Container

The can crushing container is 24 inches long by 5 inches wide. To provide adequate stiffness the sides of the container were constructed of C1010 hot rolled steel welded at the corners. Aluminum plate was used for the bottom of the container and the crushing plate to minimize corrosion from liquid remaining the cans. Openings were provided at one end of the container to allow for ejection following the crushing action. A small pneumatic cylinder was mounted in one of the openings on one side of the container. When the can is completely crushed and the crushing plate has cleared the ejection opening on its return stroke this cylinder actuates and ejects the crushed can through the opening on the opposite side of the container into the receiving bin.

The crushing plate rides on linear bearings which move on two 0.625-inch shafts attached to the crushing container. This provided for smooth movement of the plate when exposed to non-symmetrical loading during the crushing action. The crushing plate was attached to the main pneumatic cylinder using a steel insert to strengthen the attachment.

The Holding Bin

The holding bin is constructed of aluminum sheet and has dimensions of 36 x 42 x 3 inches. The holding bin includes 6 separate storage columns, each capable of storing 15 cans for a total capacity of 90 cans. The holding bin is positioned over the can crusher and dispenses one can at a time through the can chute on command from a triggering mechanism actuated by crushing plate retraction.

The Control System

The control system consists of the actuating switch, an emergency shutdown switch, and various limit switches and interlocks that ensure that individual actions occur only after precursor actions have been initiated or completed. The intent of the control system is to insure that the can crusher can only be actuated by positive action on the part of the operator, that in case of emergency the operation can be terminated immediately, and that mechanism events during the crushing cycle occur only after required precursor events have been initiated or completed.

Actuation of the start switch will initiate the mechanism which will then complete a single can crushing cycle. This insures that control of the operation is retained by the operator providing an additional level of safety for the handicapped student. It also reinforces the relationship between operator action and subsequent mechanism operation, one of the goals of the School. A limit switch limits the movement of the crushing plate to deliver a can crushed to 1-inch height. When the plate reaches this position the control system stops the movement and reverses the movement of the pneumatic cylinder driving the plate. When the crushing plate clears the ejection opening a triggering device actuates the small pneumatic cylinder ejection mechanism which ejects the crushed can into the receiving bin. When the crushing plate reaches the fully retracted position a triggering mechanism actuates, dropping another can into the crushing container from the holding bin. The crushing plate is then positioned for another cycle and the operation is terminated.

Operational Experience

The can crusher has been in operation at the School since June 1991. The students are enthusiastic about operating the crusher and school personnel have been very positive about the benefits derived. The can crusher has been an asset in the collection and sale of aluminum cans and has been an effective training device for the students, as well.

B.14 Design Project B.14

An alternate means of communication for the hearing impaired was developed pursuant to the described need by employing a vibrotactile stimulator. This device can be used to alert the wearer to respond in some prearranged manner, possibly through the use of hand signals or other. It is primarily useful for attracting a hearing impaired person located some distance away, as in the case of students on a playground that need to be attracted by a teacher. The device consists of a small receiver with an internal vibrator positioned on the hearing impaired person and a transmitter under control of the person needing to establish communication. The most important considerations in this project were to keep the cost of the device low, provide a small but durable receiver package, and gain the user's approval of the effectiveness of the communicator. The cost of the development unit was $120. This could be reduced for a production run, possibly to less than one-half. Use of a microprocessor for receiving and processing the transmitter signal would require less power, would be more precise and would weigh less. When this project was accomplished a microprocessor could not be obtained.

B.15 Design Project B.15

Based on the need described a dolly/trailer system to transport vehicles was designed. A model of the system was also constructed to verify satisfactory operation. In use the dolly was placed under the vehicle to be transported and the vehicle was raised from the ground. The dolly and vehicle were then moved up a loading ramp onto the trailer by using a wench and cable system. The system worked reasonably well but could not accommodate supercab and crew cab pickups, nor the MG Midget automobile. The pickups required a much longer dolly than was required for normal automobiles and the MG Midget was too short to fit the dolly.

The trailer/dolly system designed was a relatively simple apparatus that utilized readily available materials for fabrication, required no extraordinary skills for construction, and was able to haul almost any vehicle, damage free. Used properly, it was considered to be both simple and safe in operation.

B.16 Design Project B.16

The steps recommended for solution of this design problem are as follows:

1. Select an appropriate value for the stripping factor (R). This needs to be an iterative process to identify the lowest cost approach.

2. Select the packing material considering packing factor, cost and size (relative to tower diameter).

3. Select an optimum value for the tower pressure drop per foot of packing.

4. Compute the number of packing units (NTU).

5. Determine the allowable gas flowrate (L').

6. Determine L'/G' and use Figure 9.8 to determine the tower diameter.

7. Compute KLa and the height of the transfer unit (HTU).

8. Compute the packing height (Z).

9. Select an appropriate fan for the tower.

10. Select an appropriate water pipe diameter.

For this problem the following specifications would seem to be reasonable:

1. Height of tower = 20 ft.

2. Size of blower = 301 CFM.

3. Size of pipe = 4-inch, Schedule 40 PVC.

4. Packing material = 2-inch Berl Saddles.

5. Diameter of tower = 4.0 ft.

6. Construction cost = $6,000 to $10,000.

B.17 Design Project B.17

This project is based on the design concept that was selected for propellant handling at the Thermochemical Test Area of the Johnson Space Center, Houston, Texas.

There are basically three separate considerations that must be addressed in this project, namely:

1. Insuring that personnel outside the test complex are not exposed to toxic propellant vapor levels in excess of the TLV-TWA.

2. Insuring that intentional and unintentional spills of oxidizer and fuel can be handled in a manner such that ground and surface water is not contaminated.

3. Devising a method of operation that does not impose an unreasonable burden on test complex personnel.

The problem of exposure of nontest complex personnel to excessive propellant vapor levels was solved by constructing a boundary fence around the complex that was positioned such that no credible failure or operational procedure would result in vapor levels exceeding the allowable TLV-TWA. Atmospheric conditions were monitored and operations were restricted to conditions that would disperse any propellant vapor by the time it reached the boundary fence. The fence was posted with warning signs that provided instructions to follow in cases when visable propellant vapor clouds were released. A few significant equipment failures were necessarily eliminated from consideration in adopting this concept. A main storage tank failure resulting in dumping the total quantity of fuel or oxidizer was not considered viable since the tanks were designed to ASME Code (4:1 safety factor) and the storage tanks were unfired vessels (operational control was maintained over tank pressure increases). Propellant tank venting operations were restricted to flowrates and atmospheric conditions that did not exceed TLV-TWA's at the boundary fence.

To handle small intentional and unintentional spills in the test cells and propellant storage tank aprons a floor washdown system was provided in conjuction with a process sewer system that allowed the contaminate, diluted with copious amounts of water, to be flushed into the dedicated sewer system, then routed the contaminated water to an open pond some distance from the test facilities. The flushing system could be remotely actuated or spills could be hosed down by test personnel, properly protected. Treatment of the effluent, if required, could be accomplished at the disposal pond area by either aereating the contaminated water in the pond or adding appropriate chemicals. The water added during the flushing operation eliminated any fire or toxic vapor exposure hazard from the fuel. Oxidizer spills normally vaporized before water could be added but the water tended to wash away residual chemicals.

The concepts described above allowed operations to be conducted with wery little impact on operations. Text personnel were familiar with the propellants and how to protect themselves against exposure. Once the pretest procedures had been completed, test personnel could conduct all operations remotely from the control room with visual access to the test cell. During pre and post-test operations personnel used protective equipment that was appropriate for the system condition. When working on a system that potentially could contain propellant appropriate precautions were taken and protective equipment was worn. Atmospheric condition monitoring equipment was simple and consisted of wind direction, wind velocity, and inversion temperature instrumentation. Nomographs were developed from which safe operating conditions could be derived using this atmospheric condition data.

These facilities have been in operation for over 25 years providing outstanding test support to the space program and with an excellent environmental performance record.

B.18 Design Project B.18

Follow the precudure outlined in Example 9.1.

SOLUTION FOR EXERCISE 1.1 AUTOMOBILE PASSIVE RESTRAINT DESIGN CONCEPT

V—Passive Restraint System Description

V.1—Design Philosophy

In addressing the problem of providing an acceptable and effective passive restraint system, an appropriate design philosophy must be adopted. The driver/passenger can be restrained in the seat or can be allowed to leave the seat but have his resulting motion arrested prior to impacting the steering wheel, dashboard, windshield or other internal structure. The provision of friendly interiors can significantly mitigate the probability of an unrestrained occupant receiving fatal injury but the elimination of other restraint devices does not appear warranted. Research indicates that personal safety is greatly enhanced when the occupant is restrained in the seat during an accident. It has been documented that the fatality rate increases by 500% when a person is ejected from his seat.[14] Thus, the conceptual approach recommended is based on maximizing the probability of retaining the occupant in the seat by providing a positive automatic belt restraint, by redesigning the seat to increase the likelihood of occupant retention and provide some impact attenuation, and, finally, to insure some redundency for occupant safety by including a friendly interior with an energy absorbing collapsible steering wheel assembly.

For this system the body is considered as two main masses, upper and lower, whose actions are nearly mutually exclusive.[15] The upper mass includes the shoulder and chest region, while the lower mass includes the abdomen, pelvic and leg (knee) regions. The head/neck region is considered separately since its full mass moves independently from the upper mass. consequently, three separate restraint provisions are necessary to effectively prevent injury, viz., head/neck restraint, upper mass restraint and lower mass restraint.

V.1.1—Head/Neck Region

The human skull can withstand high impact forces without bone damage when the force is evenly distributed over a large contact area.[16] A good mitigation method will alleviate many of the severe/fatal head injuries which occur from accidents of minor/moderate severity. A good mitigation method must include even and maximal surface area distribution of head contact, and consider the force penetration characteristics of the padded materials used.[17]

V.1.2—Upper Mass Restraint

In considering the standard shoulder harness with respect to chest restraint, a load of 3000 pounds, distributed over approximately 40 square inches, can be imposed on the occupant's rib cage. While shoulder harnesses are infrequently the cause of internal injuries, clavicle fractures are fairly common since the load occurs on ribs that are not well anchored. SAE supports the contention that injuries sustained are usually due to the pressure exerted by the shoulder harness on the "soft spots." By distributing the force over a larger surface area, the chance of internal injuries and clavicle fractures can be further reduced.[18]

While the shoulder harness is relatively effective in restraining motion in the sagittal plane, it is not very effective in preventing ejection or injury in a lateral impact. The impact region of the passenger compartment is important to the magnitude of injury risk while the shoulder harness merely mitigates the degree of injury. Modifications of the lateral structure of the passenger compartment can reduce injury risk by intrusion. This can be accomplished by installing reinforcing or energy-absorbing elements in the lateral car structure.[16] SAE also suggests energy-absorbing bolsters that fit into the sides of the compartment in impact-exposed locations.[18]

V.1.3—Lower Mass Restraint

The standard seat belt with a lap belt and shoulder harness does not provide satisfactory abdominal protection. The most prevalent and severe type of impact is the head-on collision.[19] Alone, the lap belt can be potentially more dangerous than no safety belt. The lap belt can cause a person to be thrust at an acceleration far exceeding the acceleration he would have reached had he not been wearing a safety belt.[20] This motion is labeled "jack knifing." "Jack knifing" causes the chest and face to strike the steering assembly, causing numerous injuries. It is considered as one of the most critical effects of the lap belt during accidents, since many people have been severly injured or killed by this phenomena. One study showed that 16.1% of all injuries occur in the abdomen area when a seat belt was not worn, while 14.9% of all injuries occur in the abdomen area when a seat belt was worn. Abdominal injury frequency is nearly equivalent whether a lap belt is worn or not.[18] It cannot be said that the lap belt does not restrain; however, it does not reduce abdominal injury risk significantly. According to Dr. John Stapp, "the optimal seat belt includes a shoulder harness and a thigh restraint rather than the traditional shoulder harness and lap belt."

V.2—Recommended System Description

The system recommended consists of three concepts combined to form an integrated approach that will insure the safety of vehicle occupants in frontal, side and rollover collisions. This integrated approach is based on the philosophy of restraining the occupant in the seat with an automatic shoulder/thigh belt; providing an advanced seat design that increases the likelihood of the occupant being retained while providing both body protection and crash attenuation; and furnishing selected friendly interior elements, including a shock-absorbing steering wheel that complements the effectiveness of the seat and provides redundancy for occupant safety.

V.2.1—Advanced Seat Design

Severy, Blaisdell and Kerkhoff have suggested that "a basic common sense approach to motorist protection from collision trauma calls for special attention to design of the critically important structure nearest the motorist, his seat."[21] To this end, a number of different seat designs have been proposed which incorporate occupant protection concepts. One design with intuitive appeal is a wraparound, deep bucket seat with strong side wings and flanges to protect the hips. The motive behind such a design is simple: to maintain occupant position and placement while providing a safe, comfortable seat. Texas Tech University's GM Project Team agrees with this line of reasoning and proposes a novel approach to the deeply contoured bucket seat in an attempt to satisfy this criterion.

The phrase "seat-within-a-seat" aptly describes the basic design of this advanced seat concept. The overall concept is a wing-style chair (Figure 9) which provides for occupant comfort and protects against lateral movement. The seat is divided into two primary components: an inner, mobile comfort zone and an outer, winged protection shell (Figure 10). Each component represents a distinct aspect of an integrated protection system and, as such, performs separately as well as together. The inner seat is composed of a base which rests on tracks, a seat pan, and a seat back. Included in the base are tracks which guide the inner seat in and out of the outer protective shell. This movement helps create the protective wings of the outer shell thereby increasing occupant protection. The seat pan is air cushioned with a manual adjustment for maximal occupant comfort. Inflation or deflation of the cushion to a desired height or cushion effect can be achieved with the use of small air pumps on the side or front of the seat. When not inflated, the seat positions the passenger in what is considered to be the optimal restraint position (default value) by lowering him/her down into the seat. The front edge of the seat is adjustable to accommodate a range of occupant leg lengths (5th to 95th percentile). Passengers can raise or lower the pan extension according to preferred leg support. Hinges under the front edge provide a pivot point for restraint system operation allowing the back of the seat to drop but maintaining front edge height. The back of the inner seat is padded and contoured for comfort. Operation of the system requires that the inner seat back be a single unit (back and head rest) to allow for safe and effective movement of the passenger downward into the default position.

The headrest is an extension of the inner seat. A halo-contoured design with padding on the

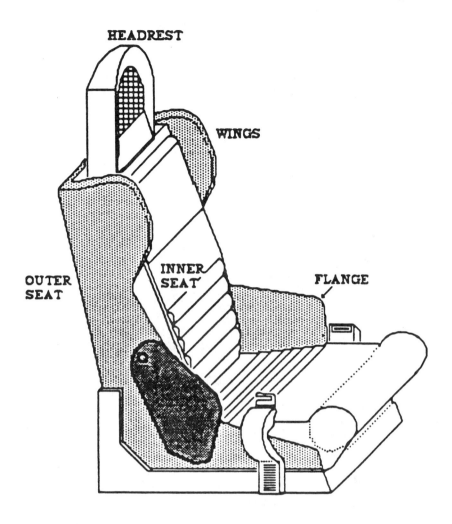

Figure 9—Advanced Seat Design Concept

top and sides and a fabric mesh inside maximizes head protection and rear visibility. The mesh fabric is attached to the seat upholstery at the bottom of the headrest as no lower halo cushion is provided. Inclusion of the headrest in the inner seat design allows the prescribed drop (2 inches) of the seat pan and seat back to occur without a concurrent reduction in occupant head protection. The objective is to provide a padded pocket for the head and neck and to reduce lateral head and neck contortion, rebound impact, and whiplash. Height and width of the headrest are based on the 5th to 95th percentile range of passenger size.

    The outer seat provides the protective framework necessary for maintaining passenger placement within the seat. It too has a base on tracks, but also includes an energy-absorbing damping system, strong side wings and flanges and is large enough to accommodate movement of the inner seat. Tracks for this base are the foundation of the force damping system. They provide the guidance for the energy-absorbing movement of the seat described later. Strong side wings and flanges are built into the outer chair for additional protection from lateral propulsion. Wings employed in

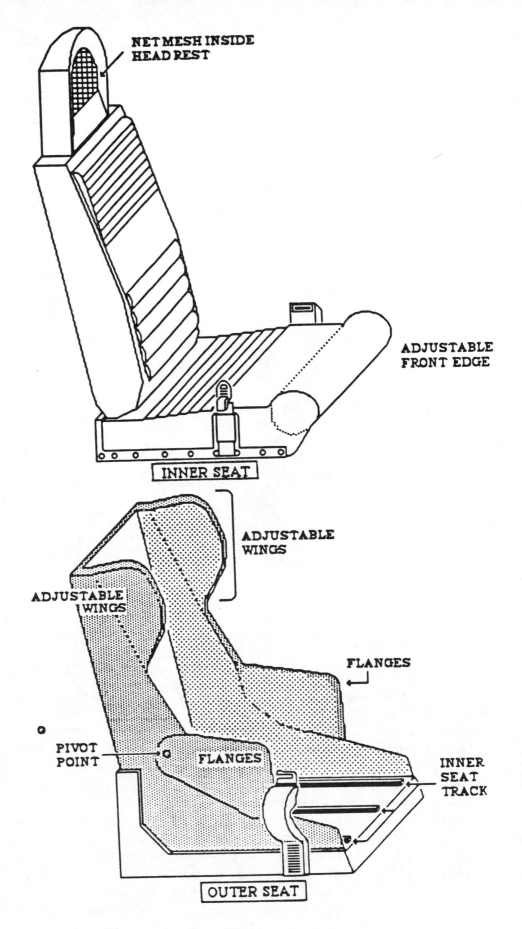

Figure 10—"Seat Within a Seat" Concept

the upper portion of the seat are fixed in height, but adjust inward and outward to accommodate both large and small passengers. Flanges, on the lower portion of the seat back, act as arm rests and help maintain occupant position in lateral impact situations. The rigid nature of these flanges provides protection from hip and pelvic area injuries common to side impacts. For two-door vehicles, as the seatback is pushed forward to access the rear seat, the flanges fold down into areas cut-out of the seat pan sides. In four-door vehicles, only the outer flange needs to be mobile to allow easy entry and exit, but having both flanges movable eliminates the need for a different seat in the four-door vehicle. In both situations, the outer flange needs to move down independently of the seat back and should operate in conjunction with either the inner seat movement mechanism or the automatic seat belt. The outer seat, non deformable to a point, acts to absorb energy from impacts (when a critical value is exceeded) by deforming, but not crushing the passenger. The entire seat system is then set on yet a third set of tracks: those normally found in vehicles and used commonly for seat adjustment (forward and backward).

The energy-absorbing mechanisms of the system are enclosed within the base of the outer chair. A set of tracks functions as the foundation for this energy-absorption system. The front ends of the tracks are curved upward (approximately 1 1/2 inches in height) while the rear ends are equipped with sturdy springs or another mechanical damping device. Given sufficient application of force to the seat, the seat travels forward (approximately 2 inches) and upward on the track to the maximum allowable height. Forward movement of the set simultaneously deflates the inner seat cushion and lowers the passenger into the seat pan. The combination of the rise on the front end and the damping device on the rear end of the tracks reduces impulses on the passenger while lowering of the seat pan (of the inner seat) lowers the moment of force which tends to thrust passengers out of the seat or cause submarining. Once forward motion has stopped, the seat is held in position by a ratchet-type device which prevents the seat from returning to the resting position. Passengers can release the seat and return it to the resting position by activating a simple catch release similar to that used in seat adjustment.

System operation can be separated into pre-operational and operational stages. In the pre-operational stage, the inner seat is positioned forward. Upper-seat back wings protrude approximately nine inches from the seat back. Inner flanges extend approximately 8.7 inches out from the seat back in their upright position while the outer flanges are lowered into their recessed position to minimize entry obstruction. The second operational stage is activated by the pre-ignition setting of the ignition switch. Activation of the ignition system engages the automatic seat belt mechanism. Outer flanges, operating in conjunction with the seat belt, move up into operational position (a height of 16.7 inches). After the seat belt is engaged, the inner seat recesses two inches into its protective outer shell. Movement of the inner seat back causes a corresponding increase in wing and flange extension resulting in two additional inches of wing and flange protection for the chest and pelvic regions. A complete table of anthropometric design considerations is presented in Tables 1 and 2. Reversal of the seat is accomplished through a return to the "engine off" position of the ignition switch. The seats return to their pre-operational positions, and the outer flanges drop to allow the passenger an unobstructed exit from the vehicle.

V.2.2—Automatic Seat Belt

The proposed automatic belt is a variation of the conventional three-point-belt system. The system includes a pressure sensor electronically linked to a motorized telescoping structure which is connected to the belt system (Figure 11). The telescoping arm functions to automatically position the three-point-belt upon the occupant when the occupant's door is closed, sufficient pressure is detected within the seat and the ignition is finally engaged. Individual pressure sensors located in every passenger seat will eliminate unnecessary belt engagement in unoccupied seats. When packages or luggage may provide adequate pressure to activate the automatic belt system, the

Measurement	Gender	5th	97th
1. Sitting Height *		32.0	37.4
2. Midshoulder Height (seated) *		21.4	26.1
3. Elbow Rest Height (seated) *		7.3	11.4
4. Thigh Clearance Height *		4.3	6.5
5. Hip Breadth (seated) *		12.8	16.3
6. Buttocks-to-Popliteal Length (seated) *		17.2	20.9
7. Head Breadth + Max Ear Protrusion **		6.5	7.4
8. Tragion to Wall Distance *	F	3.5	4.6
9. Tragion to Wall Distance	M	3.3	4.9
10. Bideltoid Breadth	F	15.1	18.1
11. Bideltoid Breadth	M	16.3	19.6
12. Bust Depth	F	8.2	10.7
13. Chest Depth	M	8.0	10.5
14. Abdominal Depth (seated)	M	7.7	10.4
15. Thigh-Thigh Breadth (seated)	F	13.3	17.0

References for Anthropometric Measurements

1-6, 8* Kodak 50/50 male and female population (22)
7** NASA #285 M-USAF Basic Trainees 1965 (23)

Table 1—Antropometric Measurements for GM Seat Design

servo motors will reverse direction when obstructions are detected, thus protecting the motor and the telescoping arm from damage.

The telescoping structure is comprised of three, curved sections of tubing of differing cross-sectional size, which curve along the same radius and which, when extended, form an arc extending from one side of the seat pan to the other. In the stowed position, the smallest section retracts within the next smallest section, etc. In its retracted state, the telescoping arm is essentially invisible, resting next to the seat pan (Figure 9).

The function of the telescoping arm is to automatically place the three-point belt upon the occupant when the ignition switch is turned on. The activation of this electrical circuit initiates the extension of the telescoping arm. The arm is extended via a flat, flexible, geared rack, which is coiled within a circular housing, and which is driven by a low-speed motor. The rack extends within the sections of the telescoping arm, and is permanently affixed to the end of the smallest, inner section. Thus, as the rack is moved by the motor, it pushes the smallest section, which in turn pulls the next smallest section, and so on. By virtue of the fact that these sections are curved along the same radius, the arm will follow an arc that is determined by this radius. This arc is such that when the arm is fully extended, the end of the smallest section will come to rest at the point where the buckle rests on the other side of the seat pan. The arc will allow an 8" thigh clearance to accommodate the 95th percentile passenger. The arm will carry both the shoulder belt and the over-the-thigh belt, simultaneously.

The latch of the three-point-belt is attached to the smallest section of the arm in such a way that when the arm approaches full extension, the latch will engage the buckle. Engagement will be facilitated by a flanged extension of the buckle. This flanged extension will function like a funnel, guiding the latch inward until it is engaged. Once the latch is fully engaged, another electrical circuit will be completed through the buckle. The completion of this second circuit will cause the telescoping arm to retract to its stowed position leaving the three-point-belt upon the occupant.

Head rest height.	37.7 in.
Head rest breadth.	11.4 in. **
Head rest wing width.	3.3 in.
Seat Back height.	26.4 in.
Seat Back breadth (top).	19.9 in.
Seat back breadth (bottom).	16.6 in.
Wing top.	26.4 in.
Wing bottom.	18.7 in.
Wing width.	9.0 in. *
Seat pan length (minimum).	17.5 in.
Seat pan length (maximum).	21.2 in.
Seat pan breadth.	17.3 in.
Flanges height.	6.8 in.
Flanges width.	8.7 in.

* widths are calculated from the inner seat before retraction
** measurement includes a 4 inch allowance to provide an unrestrained yet protective fit

Table 2—Seat Design Dimensions

The lower belt will be positioned across the upper thighs of the passenger.

The belt is disengaged manually by the occupant in the same manner as conventional belts. When the latch is disengaged, retractors pull the belt back toward the telescoping arm. Guides located on the inner side of the telescoping arm will function to guide the belt as it retracts. The latch will come to rest on a hook or connecting device such that the three-point-belt-is once again attached to the telescoping arm and is in the stowed position when the occupant next enters the vehicle.

The design simplicity and economical use of space allow the automatic belt to be easily integrated with the seat design described herein. The anchor points for the telescoping arm and the three-point-belt must be situated on the seat itself. The retractors for the belt will be mounted on the inner seat while the telescoping arm and its motor will be mounted within the outer seat (Figure 10). This arrangement will allow the occupant to be belted in securely and will eliminate the possibility of abdominal injury from belt tightening as a result of relative (inner versus outer) seat movement during a collision. If the anchors were placed on the B-pillar, side rail, and center tunnel of the car (as with conventional three-point-belts), in a collision the belt would tighten around the occupant, exerting extreme forces upon the body as the seat moves to absorb some of the impact. Also, the telescoping arm must be able to maneuver, through or around the side flanges that form the arm rests. It is proposed that the telescoping arm be mounted within the outer seat structure. The buckle and the telescoping arm would be installed in the seat structure and extend through

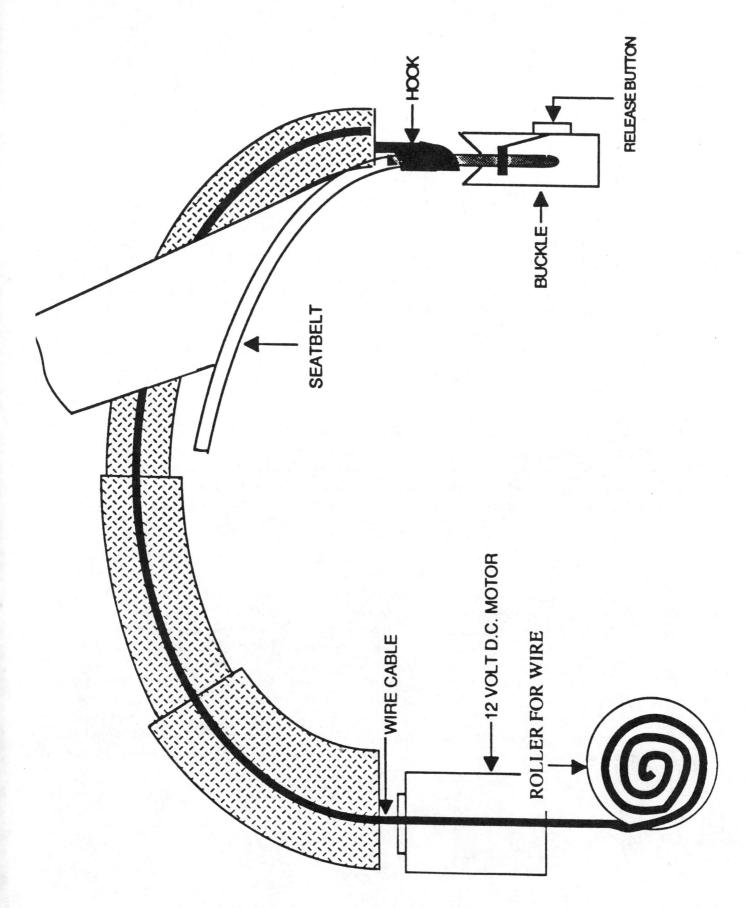

Figure 11—Passive Three-Point Seat Belt Concept

the flanges to allow the arm to extend and retract. The flanges will act as an additional guide for the telescoping arm.

V.2.3—FRIENDLY INTERIOR

"According to the laws of motion, first explained by Issac Newton 300 years ago, the occupants in a motor vehicle that suddenly stops continue to move forward at the speed prior to the accident. Their motion continues until they are brought to an abrupt halt, usually by the steering wheel, the windshield or the dashboard. For example, in a typical front-end collision at 30 mph, a 150-pound person will slam into the car's interior with the force of 4500 pounds. Put another way, the unbelted occupant of a 30- mph vehicle crash experiences g forces more than 20 times greater than those an astronaut experiences at blastoff. The unbelted occupant of a 30 mph vehicle crash hits the windshield or other interior surface of the vehicle with the same impact as a fall from a three-story building."[24]

The above highlights the hostile environment which is produced in the interior of an automobile during an accident. The tremendous forces exerted by the vehicle interior on the occupant give strong impetus for the concept of friendly interiors. The actual "passenger compartment structure can be thought of in two ways, as a strong outer shell with appropriate depths of padding on its internal surfaces, or as a single deep structure from inside to outside which deforms several inches when struck locally on the inside but which is much stiffer and does not easily crush when impacted on the outside."[25] A combination of the strong exterior coupled with appropriate levels of interior padding were utilized as the primary structural model for this concept development. The characteristics of this structural model serve to maximize energy attenuation and minimize injury severity.

V.2.3.1—Exterior Structure

In designing a strong energy attenuative exterior, individual structural elements must be examined with regard to frontal, side, and roll-over accidents. Moreover, integrity of structural elements must be maintained during these collisions if occupants are to be afforded maximum protection. Polyurethane foam material is highly successful in absorbing energy. For example, according to a National Highway and Safety Administration study "foam filled structural elements internally primed and with foam properly encapsulated, are effective in absorbing energy. The amount of energy absorbed does not vary significantly as a function of processing or environment."[26] However, it is critical that the polyurethane foam be placed in areas that will facilitate maximum energy absorption (Figure 12). Strategically placed foam filled body panels should greatly decrease the amount of collision energy transferred to the passenger compartment during a frontal collision.

When considering side impacts, two areas of the door should be examined. These areas include the location of reinforcing members and the use of padding. In a side impact, intrusion by a foreign object is a major problem. To minimize this problem the strength of the door has to be increased. Most doors that are produced today offer little or no internal support. They mostly consist of a small perimeter frame with a welded outer skin. In the past, this design may have been sufficient because the doors were big and had a large mass available to absorb energy. Today, economy is the goal and mass has to be sacrificed. Therefore, the smaller doors of today need more support to protect against intrusion.

We propose a thin-walled metal alloy tubing support that will create an inner frame for the door. This frame will be designed in such a way that no interference with the window mechanism of the door will occur. The basic shape of the frame can be seen in Figure 13. As illustrated, the tubing will run vertically and horizontally aross the structure of the door conforming to its shape. The actual diameter and spacing of the tubing will depend on the individual door design. To further reduce the injury index caused by side impacts, the interior of the door should include a sufficient amount of energy absorbing material. Hexcell aluminum honeycomb material could be

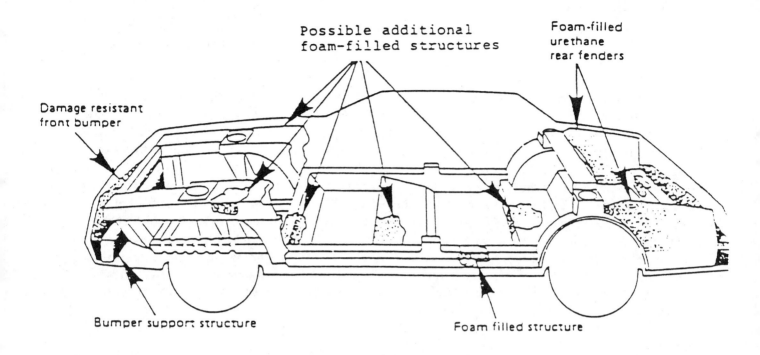

Figure 12— Foam to Structure Application

used in this application (Figure 14).

Due to the tremendous forces encountered by a car during a roll-over, the doors, roof, and door posts should be heavily reinforced. This reinforcement could be accomplished through the use of composite materials. Thermoplastic elastomers (composites) are gaining wide acceptance in the aviation and automotive industries. These materials possess superior strength-to-weight ratios. Furthermore, "factors driving this increased use are weight reduction and corrosion-protection requirements, styling, economics and company marketing strategies."[28] In an effort to maximize safety, simplify assembly, and lower tooling costs, the concept design could utilize composite materials in critical body parts. Materials considered for use include the Bexloy series of composites by DuPont, Bayflex 150 by Mobay, and Xenoy by General Electric. Both the Bexloy C-712 and the Bayflex 150 have been successfully used in the Pontiac Fiero.[29] For example, when "the National Highway Traffic Safety Administration (NHTSA) crashed a Fiero in its experimental 35 mph test series, the belted dummies inside the Fiero recorded the lowest injury numbers ever seen in the six years these tests have been run."[30] Moreover, from the roll-over standpoint, the utilization of composite doors, roof, and door posts should create an effective rollcage-like structure. (Figure 15).

V.2.3.2—Interior

A strong exterior will help greatly to reduce injuries during a collision. However, in a collision

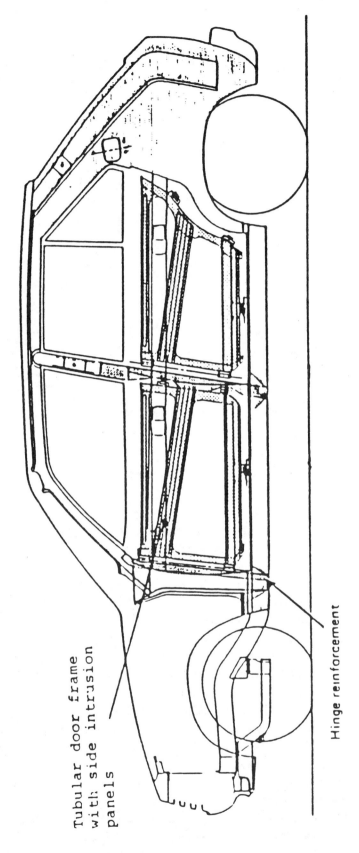

Figure 13—Door Frame Structure

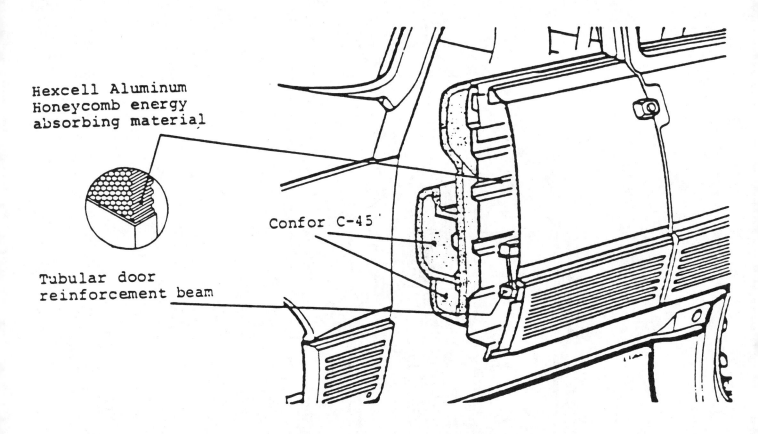

Figure 14—Inner Door Energy Absorbing System

"the belted occupant's extremities are flung about and as a result they may strike almost any part of the car interior. Their velocity of impact with the interior is largely a function of the dynamics of the impact between the car and the fixed object, commercial vehicle or other car. Energy-absorbing padding at all interior parts of the passenger compartment is needed whether or not there is intrusion into the passenger compartment."[25]. It is for this reason that a "friendly interior" has been fully incorporated in this design concept.

The purpose of the friendly interior is to provide maximum occupant protection from otherwise dangerous interior components. The friendly interior is primarily composed of a polyurethane material. The dash, door panels, and headliner should all be covered with appropriate thicknesses of polyurethane foam. In addition, two other features of the friendly interior would include redesign of the steering system and the incorporation of LOF safety glass.

A polyurethane foam similar to that used in the space shuttle could be utilized as the primary construction material for the dash and other parts of the interior. Confor C-45 is the polyurethane material which is currently being used in the space shuttle seats. This foam absorbs 85% of the total energy encountered by the shuttle astronauts during take-offs and landings. Also, when deformed rapidly in a collision the material becomes firmer, and recovery from deformation is slow so less energy is returned to the impacting body.[31] Appropriate depths of a polyurethane material (i.e. Confor C-45) could be utilized in an effort to increase the impacted area of the dash. By increasing the area of the impact during an accident, energy is dispersed over a greater portion of the body,

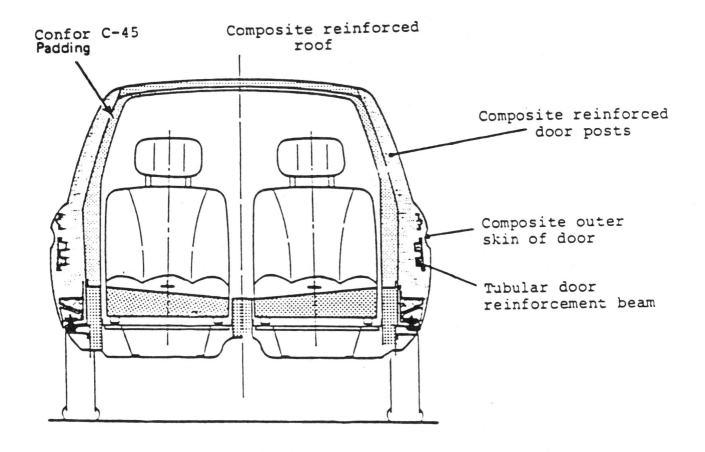

Figure 15—Roll Cage Structure

decreasing impact on any one area. We recommend a totally nonobtrusive, large surface area dash board designed to absorb and distribute loads during impact. Other objects that a driver or passenger might hit are either recessed, such as switches and handles, or are heavily padded, such as seat back, roof pillars and the roof itself.

An area that has been responsible for many serious injuries and fatalities in collisions is the windshield. In an effort to decrease the severity of windshield related injuries the concept design includes the Libbey-Owens Ford (LOF) windshield. "LOF uses a five-layer windshield consisting of the present three-layer windshield plus two inside layers. One layer is polyvinyl butyral; the other is a polyester film." The use of the LOF windshield is preferred because "five-ply safety windshields protect against decapitation and throat injuries in an accident by preventing a passenger's head from penetrating the windshield."[32]

The biggest threats to the driver in a crash are the steering wheel and steering column. Modifications can be made to the existing energy absorbing steering systems (i.e. G.M. ball-type E.A.) to improve energy absorption during the driver's impact on the steering assembly. The modifications can best be explained by considering two driver positions during a frontal collision. If the driver is aware of an impending accident, there may be sufficient time for bracing before the collision and the primary interaction of the driver with the steering assembly would, in this case, involve his extended arms. During the collision, if the driver's force through his extended arms exceeds some predetermined value, the steering column will collapse. The driver will experience a ride-down until his chest meets the expanded chest pad which absorbs remaining torso energy. The column spring causes the steering assembly to return to its initial position, with chest pad back in its original position. If the driver is unaware of an impending accident, the primary interaction will involve the chest impacting the steering assembly. As the chest strikes the chest pad, deflection due to the energy absor

ing device in the steering column at the end of the chest pad column occurs and impact is significantly attenuated.

These modifications will require component changes and additions to the ball-sleeve column steering assembly (Figure 16). The steering wheel can be considered a rigid body that transfers the impact force from the arms to the shear capsule. The force must exceed the force exerted on the column piston due to the pressurized cylinder to compress the column. As the column is compressed, the chest pad expands from the air entering the pad through a nozzle. During this process, the steering wheel rides down with the column, which loads the spring beneath the telescoping shaft. When the maximum stroke has been reached, the driver's kinetic energy has been completely absorbed. The column then begins its return to its initial position. The spring unloads causing upward acceleration of the column. This motion is damped by the suction of the air from the expanded chest pad back through the nozzle into the cylinder until the steering wheel is back to its initial position. For those cases in which the primary impact is with the chest on the steering wheel, the chest pad receives the impact, distributes the load over the pad as the column compresses down into its shock-absorbing-like element, dissipating the torso energy more effectively than a spoked steering wheel.

The basic purpose of this design is to increase the effectiveness of the existing ball-sleeve energy absorbing steering assembly through modifications that increase the energy absorption capabilities. The main feature of the modification is the re-usable, expandable chest pad. This pad incorporates a membrane stretched over an impact cushion material to cover the pad surface area. An attractive covering will be placed over the membrane to increase its attractiveness to the occupant. When activated, the air that is displaced from the pressure cylinder will expand the membrane so that it forms a dome-like surface increasing the impact surface area by 10-20% and increasing energy absorption via the air between the impact cushion and the membrane. Activation time for this device can be minimized by optimizing such parameters as the amount of air displaced, size of air passage and nozzle, average column stroke, cylinder pressure, land membrane material. The functionality of the ball-sleeve system has been proven and is currently in use in some automobiles. This system will be virtually unnoticed by the occupant except that there is an attractive pad in place of the spokes on the steering wheel. The system cannot be de-activated and requires little maintenance. The maintenance that would be required could be accomplished in the same manner as air pressure in the tires is maintained. One of the most attractive attributes of this system is that it is designed to be used over and over again without replacement cost.

Side impacts account for nearly as many life threatening injuries as head-on collisions. Therefore, appropriate levels of padding (i.e. Confor C-45) should also be incorporated into the door design. Moreover, "padding of the interior surface of the doors should be designed so that it is effective even if the structure to which it is attached is deformed" (see Figure 14). Proper placement of the door's padding is critical. The two primary areas which must be protected during a side collision are the pelvic and thoracic regions (see Figure 17). It was found in a study of 48 lateral automotive accidents that fractures of the acetabulum with intrapelvic protrusion of the hip is characteristic. The vehicle door produced the most injuries in such cases. Therefore, the need for improvement of the door structure internally and externally is essential and must not be

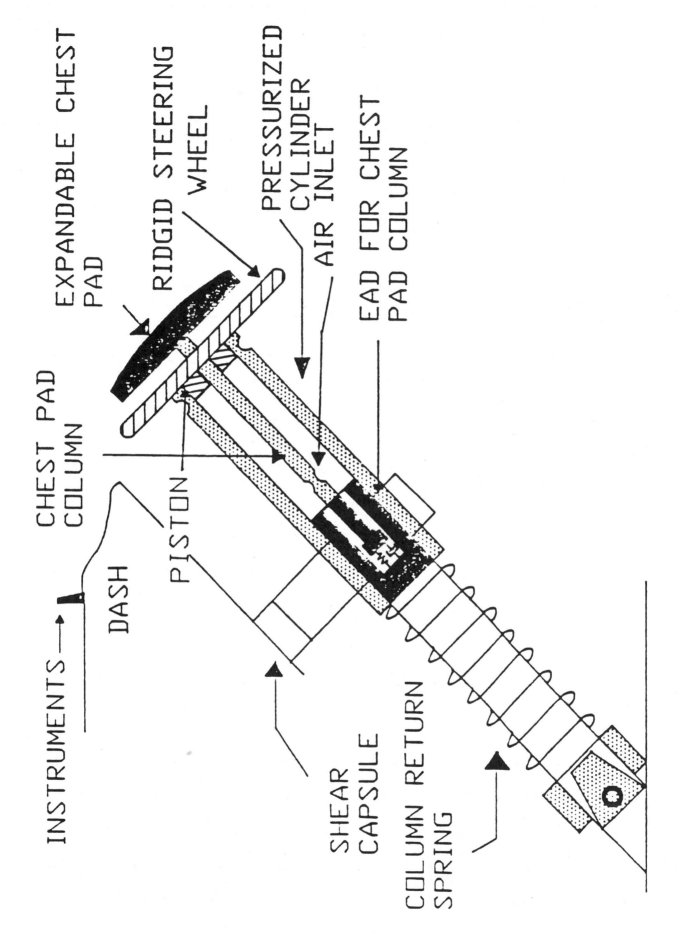

Figure 16—Energy Absorbing Steering System

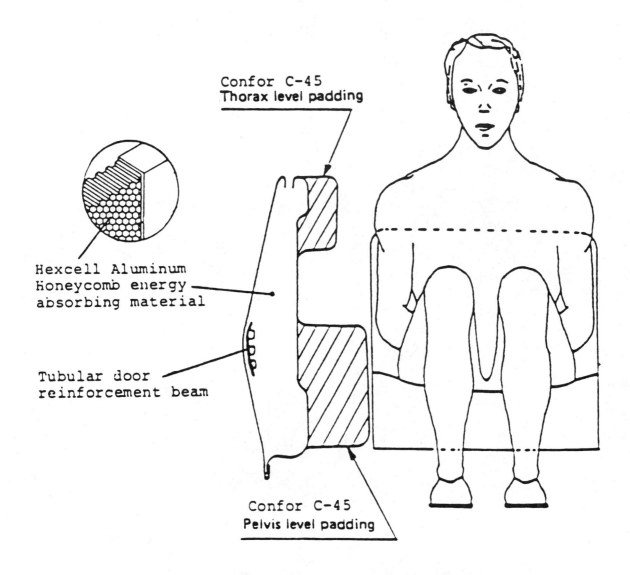

Figure 17—Thorax and Pelvis Level Padding

overlooked.

In an effort to insure occupant protection in a roll-over, a polyurethane pad (Confor C-45) should also be located under the cloth headliner (see Figure 15). This pad should be thick enough to provide adequate protection, without depleting valuable headroom; thereby adversely affecting customer acceptance.

In summary, the utilization of the strong exterior coupled with appropriate levels of interior padding has many positive features. The hexcell aluminum honeycomb reinforcement and foam-filled body panels can greatly reduce the collision energy transferred to the passenger compartment. The strategic placement of polyurethane padding (Confor C-45), intrusion panels, and the utilization of LOF safety glass create an integrated friendly interior. This friendly interior serves to provide maximum occupant protection from otherwise dangerous interior components.

VI—PASSIVE RESTRAINT SYSTEM EVALUATION

The automatic belt, in conjunction with the advanced seat and friendly interior, represents an integration of three concepts so that the integrated design satisfies the seven technical evaluation criteria established by the Request for Proposal (RFP). Although each of the three concepts exhibit individual strengths, none are totally effective by themselves, hence the integrated design. An evaluation of the system with regard to the seven criteria is outlined below.

VI.1—Functionality

The proposed system is functional in that it will meet and/or exceed the requirements set forth by CFR 49.[6] The three-point-belt has been shown to be a superior restraint device. A noteworthy aspect of the proposed belt system is the inclusion of an over-the-thigh lap belt. This belt arrangement will reduce the incidence of submarining as well as improve the ability of the system to retain the occupant during a rollover. The wings and flanges of the advanced seat will effectively reduce lateral displacement of the occupant during side impact crashes. Furthermore, the seat's force-damping base will act to increase deceleration time, thereby reducing the severity of the forces acting upon the occupant in frontal collisions. The elevation of the seat as it moves on the rails at impact, and the lowering of the body as the air cushion deflates effectively eliminates submarining and places the occupant in a much more favorable position to absorb the attendant deceleration forces.

The friendly interior provisions complement the automatic belt and advanced seat by providing increased protection for passenger compartment intrusion and offer redundant protection for the occupant if ejected from the seat. The shock-absorbing steering assembly protects the driver against injury from impact in cases where the automatic belt has been disengaged.

VI.2—Engineering Feasibility

The automatic belt and the advanced seat work together to ensure overall system performance. The inner seat, to which the belt and buckle are attached, must align with the outer seat and its telescoping arm so that the arm engages consistently and accurately with the buckle. This critical design area will be evaluated during Phase II by consideration of alternative models and options. The materials and technology for producing a structurally sound and accurate system exist, and therefore the passive restraint system proposed is considered feasible.

The collapsible steering wheel/column requires some development effort in subsystem dynamics. The subsystem must be designed to be essentially maintenance free and the chest pad must be highly durable. No other engineering feasibility problems are foreseen in the proposed system.

VI.3—Innovativeness

The restraint system proposed incorporates an appropriate mix of proven technology and new concepts. This system represents an integration of the merits of several concepts, each marginally effective when employed by themselves. The seat is an improvement over the seats available in most vehicles today as it provides superior lateral impact protection, reduces deceleration forces, and offers increased occupant retention capability.

The belt design differs from systems in existence in that the telescoping arm is not visible in its retracted state. This is considered to be an improvement over current automatic belt systems which rely on the use of retracting cables and tracks. Also, these systems often inhibit ingress and egress of the occupant. The collapsible steering wheel/column is unique in that the collapsing force inflates the chest pad, providing additional driver crash protection.

VI.4—Customer Acceptance

An informal market survey was conducted in order to determine public acceptance of passive restraint systems. Respondents, ranging in age from 18-52 years, rated their overall opinion on passive restraint as favorable (mean rating = 7.33 on a scale from 1 to 10). Respondents also indicated how much they would be willing to spend to have the system in their car. Amounts and corresponding percentage of respondents were: $50-$100 (14.4%), $100-$500 (38%), $500-$1000 (21.4%), and $1000+(26.2%). These results indicate the trend toward public acceptance of passive restraint. Although this was a very limited survey, on the basis of these results our design concept appears to be well received by consumers.

VI.5—Ease of Use

By virtue of the fact that this is a passive system, requiring no action on the part of the occupant, it should be quite easy to use. Furthermore, the system will be comfortable and, therefore should facilitate customer acceptance. System maintenance should be restricted to the automatic belt and collapsible steering wheel/column and emphasis in Phase II will be placed on simplifying the design and maximizing the reliability of these sub-systems.

VI.6—Cost

Installation of this system will be expensive. However, if the system is evaluated on the basis of cost effectiveness it should be considered favorably. In addition, the proposed system incorporates features associated with occupant comfort (advanced seat) and vehicle crash survivability (exterior structural changes) which should be considered in determining the system cost directly attributable to occupant safety.

VI.7—Manufacturability

The materials and technology necessary for the production of the components that comprise the system exist and are available. With moderate modifications to the design of the automobile and the manufacturing facility, this system could be manufactured with relative ease.